Industrial Mixing Fundamentals With Applications

Elmer L. Gaden Jr., Series Editor

Gary B. Tatterson, Volume Editor
Richard V. Calabrese and W. Roy Penney
Volume Co-Editors

André Bakker	Paul A. Gillis	Kenji Nakamura
R. Eric Berson	Hanna Gladki	Takahide Nouzawa
Vinayak D. Bhat	P. Guichardon	Kohei Ogawa
J. Borth	I.S. Hamill	N.T. Padial
Pierre J. Carreau	Thomas R. Hanley	Tushar Pattni
Jianya Cheng	I.R. Hawkins	W. Roy Penney
Robert S. Cherry	Matthew Hazleton	Anders Rasmuson
Raj P. Chhabra	I.P. Jones	T.C. Scott
Naimul H. Chowdhury	B.A. Kashiwa	Takeo Shiojima
D.W. DePaoli	Robert D. Knecht	B.A. Splawski
N. Devanathan	S. Kumar	Stevan P. Stiefvater
P. Drtina	Chiaki Kuroda	F. Streiff
M.P. Dudukovic	E. Lang	Claes Sturesson
L. Falk	S.M. Lo	C. Kurt Svihla
Julian B. Fasano	Richard Long	Hans Theliander
J.Q. Feng	Lazaro J. Mandel	Kuochen Tsai
K. Fontenot	Kevin J. Myers	C. Tsouris
M.C. Fournier	Yoichi Nagase	W.B. Vanderheyden
Rodney O. Fox		J. Villermaux
Tim W. Gambrel		S.L. Yarbro

AIChE Staff
Maura N. Mullen, Managing Editor; Julie A. McBride, Editorial Assistant
Cover Design: Joseph A. Roseti

Inquiries regarding the publication of Symposium Series Volumes should be directed to:

Mark Rosenzweig, Editor-in-Chief
American Institute of Chemical Engineers, 345 E 47 St., New York, N.Y. 10017
(212) 705-7576 • FAX: (212) 705-7812

AIChE Symposium Series

American Institute of Chemical Engineers

© 1995
American Institute of Chemical Engineers (AIChE)
345 E 47 Street
New York. N.Y. 10017

AIChE shall not be responsible for statements or opinions advanced in their papers or printed in their publications.

Library of Congress Cataloging-in-Publication Data

Industrial mixing fundamentals with applications / Gary B. Tatterson,
    volume editor, Richard V. Calabrese and W. Roy Penney, volume co-editors
    p.   cm. -- (AIChE symposium series, ISSN 0065-8812; no 305, v. 91)
    Includes index.
    ISBN 0-8169-0668-8
    1. Mixing--Congresses. I. Tatterson, Gary B. II. Calabrese, Richard, 1946-
       III. Penney, W. Roy IV. Series: AIChE symposium series no. 305
    TP156.M5152   1995
    660' .284292--dc20
                                                                     95-9510
                                                                                                      CIP

    All rights reserved whether the whole or part of the material is concerned, specifically those of translation, reprinting, re-use of illustrations, broadcasting, electronic networks, reproduction by photocopying machine or similar means, and storage of data in banks.
    Authorization to photocopy items for internal use, or the internal or personal use of specific clients, is granted by AIChE for libraries and other users registered with the Copyright Clearance Center Inc., provided that the $3.50 per copy is paid directly to CCC, 222 Rosewood Drive, Danvers, MA 01923. This consent does not extend to copying for general distribution, for advertising, or promotional purposes, for inclusion in a publication, or for resale.
    Articles published before 1978 are subject to the same copyright conditions and the fee is $3.50 for each article. AIChE Symposium Series fee code: 0065-8812/1995.

Foreword

Mixing is an essential element of the infrastructure of the chemical, petrochemical and biochemical industries. Because of the economic significance of mixing and contacting, these industries are in need of better characterization of mixing and multiphase processing. It is with this purpose that this symposium was developed.

This publication is sponsored by the North American Mixing Forum.

Gary B. Tatterson, *Volume Editor*
Department of Chemical Engineering
North Carolina A & T State University
Greensboro, NC 27411

Richard V. Calabrese, *Volume Co-Editor*
Department of Chemical Engineering
University of Maryland
College Park, MD 20742

W. Roy Penney, *Volume Co-Editor*
Department of Chemical Engineering
University of Arkansas
Fayetteville, AR 72701

Organizing Committee

Piero M. Armenante, New Jersey Institute of Technology
André Bakker, Chemineer, Inc.
David Bigio, University of Maryland
Richard V. Calabrese, University of Maryland
Jeffrey Chalmers, The Ohio State University
Deepak Doraiswamy, DuPont Experimental Station
Paul M. Kubera, Lightnin
W. Roy Penney, University of Arkansas
Steve Ruszkowski, BHR Group Ltd.
John M. Smith, University of Surrey
Gary B. Tatterson, North Carolina A&T State University
James M. Wallace, University of Maryland
Nam S. Wang, University of Maryland
Jack Zakin, The Ohio State University

Officers of the North American Mixing Forum

Past President: James Y. Oldshue
President: E. Bruce Nauman
Vice President: Art W. Etchells
Treasurer: Richard V. Calabrese
Secretary: Piero M. Armenante

CONTENTS

Foreword iii

Analysis of Mixing in the Sulzer SMV Mixer by Numerical Simulation
E. Lang, P. Drtina, J. Borth and F. Streiff 1

Numerical Simulation and Experimental Verification of the Gas-Liquid Flow in Bubble Columns
S. Kumar, W.B. Vanderheyden, N. Devanathan, N.T. Padial, M.P. Dudukovic and B.A. Kashiwa 11

Hydrodynamic Analysis of a Two-Phase Tubular Reactor
S.L. Yarbro and Richard Long 20

PDF Modeling of Turbulent Mixing and Chemical Reactions in a Tubular Jet Reactor
Kuochen Tsai and Rodney O. Fox 31

A Comparison of Experimental Data and Numerical Simulation of Mixing in Forced-Circulation Evaporative Crystallizers
Paul A. Gillis and Tim W. Gambrel 39

A Relationship Between Grinding in Fluidized-Bed Units as a Function of Reynolds Number
Robert D. Knecht, Matthew Hazleton and Stevan P. Stiefvater 45

Electrostatic Spraying of Gases into Liquids
C. Tsouris, D.W. DePaoli, J.Q. Feng and T.C. Scott 52

Mass Transfer in a Laminar Rippling Film in a Conical Centrifugal Film Reactor
Part I: Film Thickness Profile and Velocity Profile
Richard Long and Tushar Pattni 61

Mass Transfer in a Laminar Rippling Film in a Conical Centrifugal Film Reactor
Part II: Stability Analysis for the Film Flow
Richard Long and Tushar Pattni 70

Mass Transfer in a Laminar Rippling Film in a Conical Centrifugal Film Reactor
Part III: Enhanced Mass Transfer in Ripple Flow
Richard Long and Tushar Pattni 80

Semi-Direct Simulation of Flow in Turbulent Transition in a Vessel with Paddle Impeller
Yoichi Nagase, Kenji Nakamura, Takahide Nouzawa and Takeo Shiojima 88

A New Scale-up Rule and Evaluation of Traditional Rules from a Viewpoint of Energy Spectrum Function
Kohei Ogawa and Chiaki Kuroda 95

An Experimental (LDA) and Numerical Study of the Turbulent Flow Behavior in the
Near Wall and Bottom Regions in an Axially Stirred Vessel
Claes Sturesson, Hans Theliander and Anders Rasmuson 102

On the Effect of Wall and Bottom Clearance on Mixing of Viscoelastic Fluids
Jianya Cheng, Pierre J. Carreau and Raj P. Chhabra 115

Study of Micromixing in a Liquid-Solid Suspension in a Stirred Reactor
P. Guichardon, L. Falk, M.C. Fournier and J. Villermaux 123

An Experimental Investigation of Solids Suspension at High Solids Loadings in Mechanically
Agitated Vessels
Naimul H. Chowdhury, W. Roy Penney, Kevin J. Myers and Julian B. Fasano 131

Simulation and Experimental Verification of Liquid-Solid Agitation Performance
Kevin J. Myers, André Bakker and Julian Fasano 139

Power Dissipation, Thrust Force and Average Shear Stress in the Mixing Tank with a
Free Jet Agitator
Hanna Gladki 146

The Application of CFDS-FLOW3D to Single and Multi-Phase Flows in Mixing Vessels
I.S. Hamill, I.R. Hawkins, I.P. Jones, S.M. Lo, B.A. Splawski and K. Fontenot 150

Gas-Liquid Mixing and Mass Transfer in Tall Tanks
C. Kurt Svihla, R. Eric Berson and Thomas R. Hanley 161

Hydrodynamic Shear Stress Effects on the Actin Cytoskeleton and Energy Status of Cultured
Epithelial Cells
Vinayak D. Bhat, Robert S. Cherry and Lazaro J. Mandel 166

Index 178

Analysis of Mixing in the Sulzer SMV Mixer by Numerical Simulation

E. Lang, P. Drtina, J. Borth
Sulzer Innotec AG, Fluid Dynamics Laboratory, 8401 Winterthur, Switzerland

and

F. Streiff
Sulzer Chemtech AG, 8401 Winterthur, Switzerland

Introduction

Static mixers are used in various processes which include mixing, dispersion, polymerisation and chemical reaction. The flow regime can be either laminar or turbulent. Static mixers obtain their energy from the pressure drop over the mixing element. The devices usually guide some part of the fluid flow transversly across the pipe or channel to achieve a homogenisation of certain fluid properties. An overview about the various static mixer types is given by Pahl and Muschelknautz [1].

To date static mixers have only been developed and tested by experimental techniques. Due to the availability of CFD programs and powerful computers it is now possible to calculate the fluid flow and the mixing process in a static mixer. This has an impact on the development technique for static mixers. The numerical simulation provides more insight into the flow and mixing processes due to the data being available in the whole flow field. In experiments data is usually only obtained in a small part of the entire domain of interest. The behaviour of the fluid in the rest of the domain has to be extrapolated. Information about the mixing process can generally be obtained faster, and with less cost by numerical simulation.

The geometry of static mixers is normally fairly simple. Nevertheless the grid generation for the numerical simulation can be rather cumbersome and time consuming. Sometimes it is not necessary to calculate the flow field in the entire mixer. In these cases the flow field and the mixing process are only computed in the relevant part.

In the next section the computation method is briefly described. In the third section we report results of the examination of the flow field and the mixing process in the Sulzer SMV mixer. The application of a SMV mixer in a DeNOx facility is described and the numerical results are compared to experimental data in the fourth section. This is followed by a summary and the conclusions.

The governing equations

The equations for conservation of mass, momentum and energy must be solved to calculate the mixing process in a static mixer. We consider the fluid as being incompressible and steady. The equations to solve are

$$\frac{\partial}{\partial x_j}(u_j) = 0 \tag{1}$$

for mass conservation,

$$\rho \frac{\partial}{\partial x_j}(u_j u_i) = -\frac{\partial P}{\partial x_i} - \frac{\partial \tau_{ij}}{\partial x_j} + S_{ui} \quad (2)$$

for momentum conservation and

$$-\rho \frac{\partial (u_j H)}{\partial x} = \frac{\partial q_j}{\partial x_j} + S_E \quad (3)$$

with

$$H = h + \frac{u_i u_i}{2}$$

for energy conservation. The molecular fluxes τ_{ij} and q_j are expressed in terms of velocity and temperature gradients using Stokes' and Fourier's law

$$\tau_{ij} = -\mu(\frac{\partial u_i}{\partial x_j} + \frac{\partial u_j}{\partial x_i}) + \frac{2}{3}\mu \frac{\partial u_k}{\partial x_k}\delta_{ij} \quad (4)$$

$$q_j = -\lambda \frac{\partial T}{\partial x_j}$$

The program TASCflow of Advanced Scientific Computing Ltd was used to computed the fluid flow and the mixing process. The program solves the time averaged equations by a finite volume scheme. The equations are therefore converted to a mean form, by a time average process. Each variable is decomposed into a mean and fluctuating part

$$u_i = U_i + u'_i \quad (6)$$

With this substitution, the equations reduce to the following form

$$\frac{\partial (U_j)}{\partial x_j} = 0 \quad (7)$$

for the continuity equation,

$$\rho \frac{\partial (U_j U_i)}{\partial x_j} = -\frac{\partial P}{\partial x_i} - \frac{\partial}{\partial x_j}(\overline{\tau}_{ji} + \rho \overline{u'_i u'_j}) \quad (8)$$

for the momentum equation and

$$\rho \frac{\partial (U_j \overline{H})}{\partial x_j} = -\frac{\partial}{\partial x_j}(q_j + \rho \overline{h' u'_j}) \quad (9)$$

for the energy equation. Equations (8) and (9) contain terms that are not expressible in terms of mean flow variables. The well known k-ε model is used to evaluate the Reynolds stress terms $\overline{\tau}_{ji}$, $\rho \overline{u'_i u'_j}$ and the turbulent heat transfer terms $\rho \overline{h' u'_j}$ (see Launder and Spalding [6]).

Detailed information how the equations are solved by a finite volume scheme can be found in the theory section of the TASCflow manual [2].

The flow field and the mixing process in the static Sulzer SMV mixer

To study the basic phenomena of the mixing process, the flow field in a section of the SMV mixer which represents the typical flow behaviour and the mixing process of the whole mixer was computed. Because of the regular periodic structure of the SMV mixer the flow and the mixing process of that section are repeated periodically in the whole mixer. All characteristic phenomena can be analysed in this subregion.

Grid and boundary conditions

The static mixer SMV consists of several corrugated blades which are stacked on each other with alternating directions to the main flow direction. The grid has been generated for a subregion which comprises the basic geometric structure of the mixer, and consists of 65,000 nodes.

To cover the entire region of interest, that is both the SMV mixer and wake region, the domain was also split into three subregions in the streamwise direction. The computations of the flow field were carried out sequentially. A plot of the computational mesh is shown in figure 1.

To examine the mixing process the inlet region was divided into two parts each comprising a mixer layer. The inlet temperature of the upper layer was $T_1=300K$, whereas the temperature of the lower layer was $T_2=301K$. In both inlet regions the velocity was 15m/s and the density $\rho=1.2kg/m^3$. The turbulent properties were unknown and therefore had to be estimated. A reasonable guess is $I=5\%$ for the turbulence intensity and for the turbulent length scale $L_t=d_h/10$ with the characteristic length $d_h=1.5m$. The turbulence intensity was later changed to investigate the effect of the turbulence inlet conditions on the mixing performance of the SMV mixer.

Periodic boundary conditions were applied to the surfaces parallel to the main flow direction. This leads to an infinite flow field in the two directions perpendicular to the main flow direction, whereby the flow pattern is repeated with the same geometric period of the subregion.

The simulations in the streamwise direction were performed sequentially. The distribution of the velocity components, the turbulence properties and the temperatures at the outlet of each section were used as the inlet boundary conditions of the next section to be computed. Hence it has been assumed that the downstream fluid flow does not influence the flow upstream, or that this influence is so small as to be negligible.

Results of the simulations

The Reynolds number calculated with the characteristic length and the main flow velocity is $Re=426,000$. The mixing process is analysed by studying the velocity field and the temperature distribution in cross sections at right angle to the main flow direction in the SMV mixer itself and in the wake region.

An analysis of the flow field shows that the corrugated sheets divert the fluid in the direction of the folded edges. That means that one layer diverts the fluid to the right and the adjacent mixer layer to the left. This behaviour is responsible for a global homogenisation of the fluid properties. In figure 2 the velocity components in a plane perpendicular to the main flow direction are shown for a location in the SMV mixer and another in the mixer wake. The velocity field clearly shows that the points, where two adjacent mixer layers touch each other, are the starting points of streamwise vortices. The vortices which are generated by the first rows of touching points travel downstream until they disappear due to viscous diffusion or are merged with the new vortices generated by the second rows of touching points. The vortices generated by the second rows can be found for a long distance downstream in the wake of the mixer.

The flow pattern described above governs the mixing process. The mixing in a SMV mixer is mainly due to these vortices and not due to increased turbulent diffusion generated by the mixer structure. No significant difference was found if the turbulence intensity at the inlet is changed from $I=5\%$ to $I=1\%$. A considerable part of the mixing process occurs in the wake of the mixer.

A contour plot of the temperature in figure 3 for the cross section at the mixer outlet also shows the mixing behaviour induced by the generated vortices.

The intensity of segregation was calculated for the mixing process, and is plotted as a function of the dimensionless mixer length in figure 4. The intensity of segregation was calculated to be

$$s = \frac{\sigma^2}{\sigma_0^2},$$

where

$$\sigma = \frac{\sum_{i=1}^{n}(T_i - \bar{T})}{n-1}.$$

All 2,275 grid points in a cross section were used to calculate the variation σ.

The intensity of segregation decreases rapidly in the downstream direction. This shows that the mixing process is quite intensive due to the arrays of the vortices. The fluid flow and the mixing process were also calculated for a SMV mixer in a channel with the same geometrical characteristics. This computation showed that the presence of channel walls slows down the mixing performance. The vortices can still be found in the wake region but they are not as regularly distributed and are distorted in shape as compared to the ideal case.

Velocity measurements and comparison with numerical results

An experiment with a structure similar to the SMV mixer was carried out to validate the numerical results of the velocity field. The details about the experiment can be found in Zhang [3]. The characteristic length of the structure in the experiment was much smaller ($d_h=0.015m$) than the characteristic length of a SMV mixer in a DeNOx facility ($d_h=1.52m$). The experimental set-up is shown in figure 5. The test section consists of several corrugated blades which are stacked on each other in the same way as the SMV mixer. Here the corrugated blades are much longer in the streamwise direction as in the case of the SMV mixer. The same flow field is generated in the structure as in the SMV mixer.

The velocity was measured with LDV between layer three and four, in a plane which is shown in figure 6. The optical axis of the LDV probe is also shown in the cross section B and C. Two velocity components were measured in the plane perpendicular to the optical axis along the line shown in detail D of cross section B.

The fluid flow in the structure was computed. The calculated velocities were projected onto the direction of the measured velocity components. The comparison for the two components is shown in figure 7.

The curves of the measured and the computed velocity for the two components considered, have the same shape, but the computed curve is shifted some distance to the left. The location, where the velocity was measured, relative to the structure could not be determined accurately. This is the reason, the two curves run parallel but a certain distance apart. Taking this into account the agreement is very good.

The flow field in part of a DeNOx facility with a SMV mixer

During the last decade the reduction of chemical compounds such as NO_X and SO_2 in flue gas emissions from conventional power plants and waste incinerators has become an issue due to stringent environmental protection regulations. The removal of nitrogen compounds is carried out by catalytic processes in a so called DeNOx facility. The catalyst requires certain conditions for velocity, temperature and chemical species concentration in the flue gas stream. The distribution of the flow variables should be as homogeneous as possible to get a uniform load and to achieve optimal operating conditions. The flue has therefore, in general, to be altered by appropriate means in order to satisfy all required conditions. This is usually achieved by employing static mixers, baffles and a number of guide vanes as described by Drtina et al. [4].

The computation of the fluid field and the concentration distribution in an entire DeNOx facility was not possible with our computer resources to date. Therefore only a section comprising the triple inlet, the first mixer, the elbow and a channel section was considered.

Grid and boundary conditions

The grid for this computation consists of three parts. Part one comprises the section from the inlet to the static mixer, part two the SMV mixer and part three the channel section with elbow to a plane some distance downstream of the ammonia injection. The grid of the SMV mixer consists of 89,375 nodes. The inlet part is modelled by 4,200 and the channel section by 12,000 grid points. A plot with the gridlines of the surface of the domain can be found in figure 8. The three grids with different node densities are attached to each other. This allows computations of complicated geometries because only the complex section must be modelled with high resolution. The mixer used here has a length $L_M=3.4m$, height $H_M=3.0m$ and depth $B_M=15.0m$.

The main flow velocity at the inlet was $v=26.9m/s$. The turbulence intensity was chosen to be $I=5\%$ and the turbulence length scale $L_t=0.5m$. To investigate the temperature equalisation the temperature was set to $T_m=400K$ in the middle section of the inlet and to $T_r = T_l = 300K$ for the right and left section. The channel walls were assumed to be adiabatic. The outlet boundary conditions prescribed were of constant static pressure across the outlet, and with the streamwise gradients of all variables taken to be zero.

Numerical results and comparison with measurement

Figure 9 shows the velocity distribution in planes perpendicular to the main flow direction. At the outlet of the computational domain in the vicinity of the ammonia injection plane strong velocity gradients can be observed. These gradients are caused by two regions of low velocity. This is the result of the secondary flow in the bend and the burners which divert the main flow to the opposite wall.

The temperature distributions in the same planes are plotted in figure 10. The mixing behaviour in the SMV mixer can easily be recognised. The hot flue gas which enters the facility through the middle inlet section is diverted to the right and to the left according to the orientation of the four mixer layers. This can be observed in the first plane behind the mixer. In the wake region the flue gas is mixed due to vortices which are generated as described in the previous section. This process is supported by the secondary flow caused by the bend of the channel.

Computations have also been carried out without SMV mixer and the result can be seen in figure 11. The effect of the mixer is obvious if both figures are compared. A much more uniform temperature distribution is achieved, and the velocity differences are smoothed out.

The results obtained by numerical simulation of the fluid flow and the mixing process were compared to measurements obtained from an experiment with a scale model. Details of the measurement are described in Streiff et al. [5]. In figure 12 the measured deviation of the local value to the mean value for species concentration is compared with the deviation of the computed temperature from the mean value along three lines. The positions of the lines are given in figure 8. Satisfactory agreement is obtained near the wall opposite the burners for the velocities and the temperature distribution, whereas just above the burners differences are found. This is caused by the rather coarse grid of this section. A higher grid resolution should lead to a better agreement.

Summary and conclusion

The flow field in an SMV mixer with and without endwall effects was studied. The analysis showed that the mixer process was governed by arrays of vortices. A considerable part of the mixing occurs in the wake region of the mixer.

The computed and measured velocity in a structure similar to the SMVmixer were compared, and good agreement was found.

An industrial application of the mixing process in part of a DeNOx facility was investigated. The simulation showed that the temperature maldistribution was equalised by the mixing process initiated by the SMV mixer. The computed results were compared to experimental data, and satisfactory agreement was found.

The investigation showed that numerical simulations can be used to compute the fluid flow and the mixing process in static mixers. A numerical simulation can give more insight into the flow and mixing of such facilities. This can lead to a better design and a faster design process. This in turn leads to devices with higher efficiencies, and reduced operating costs.

List of Symbols

d_h	Characteristic length
I	Turbulence intensity
L_t	Turbulent length scale
n	Number of data points
Re	Reynolds number
s	Intensity of segregation
T	Temperature
$\bar{T}$	Mean temperature
v	Velocity
ρ	Density
σ	Variation

Subscripts

l	Left
r	right
0	Inlet

References

[1] PAHL, M.H., AND MUSCHELKNAUTZ, E., Static Mixers and their Applications, Int. Chemical Eng., Vol. 22, No. 2, p. 197–205, (1982).

[2] **TASCflow User Manual**, Advanced Scientific Computing Ltd., Waterloo Canada (1994).

[3] ZHANG, Z., Einsatzmöglichkeit des LDA/PDA-Systems in Trennkolonnen zur Messung der Strömungsgeschwindigkeit und Wassertropfengröße, Internal Report, Sulzer Innotec AG, (1993).

[4] DRTINA, P., LANG, E., FLEISCHLI, M. AND STREIFF, F., Optimization of a DeNOx Facility by Numerical Simulation, UIT XI Congresso Nazionale sulla Trasmissione del Calore, Milan, (1993).

[5] STREIFF, F.A., FLEISCHLI, M., DRTINA, P. AND LANG, E., "Optimierung von Entstickungsreaktoren durch statische Mischsysteme und numerische Strömungsberechnung", VGB Kraftwerkstechnik (1993).

[6] LAUNDER, B.E. AND SPALDING, D.B., "The Numerical Computation of Turbulent Flows", Comp. Meth. Appl. Mech. Eng. Vol. 3, pp. 269 – 289 (1974).

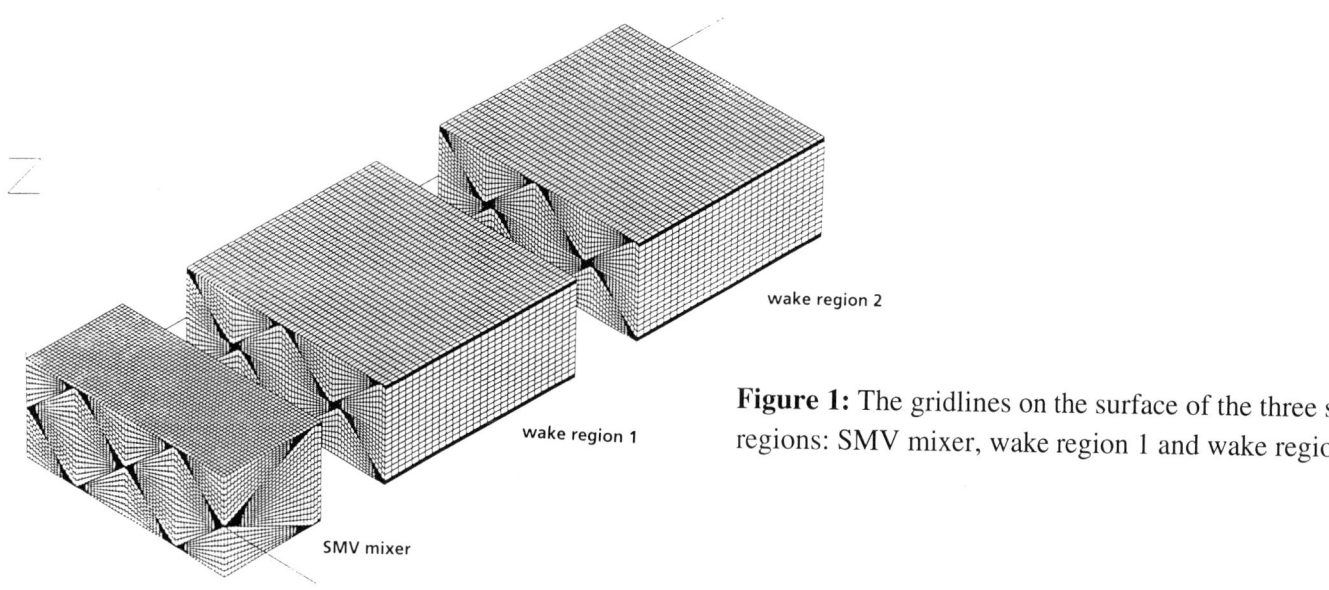

Figure 1: The gridlines on the surface of the three sub-regions: SMV mixer, wake region 1 and wake region 2.

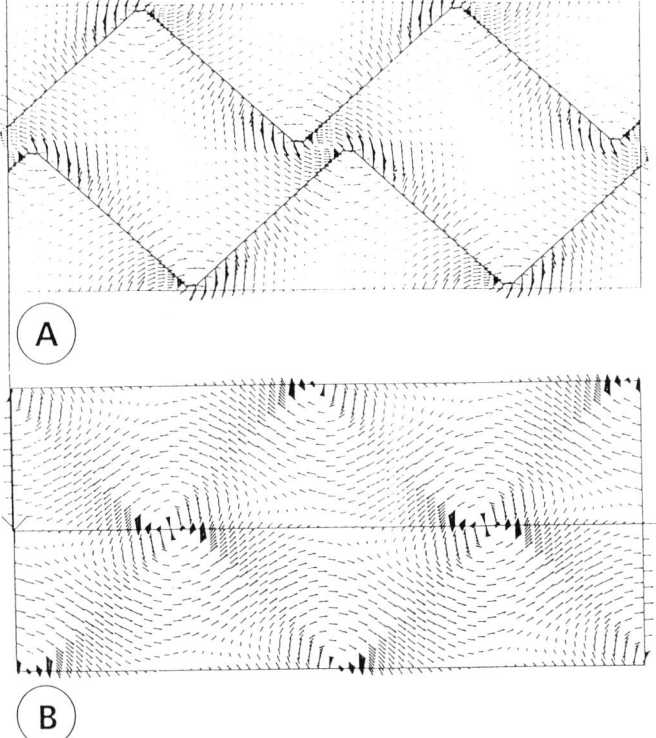

Figure 2: Velocity vectors in planes perpendicular to the main flow direction in the SMV mixer (A) and in the wake region (B). Endwall effects not included.

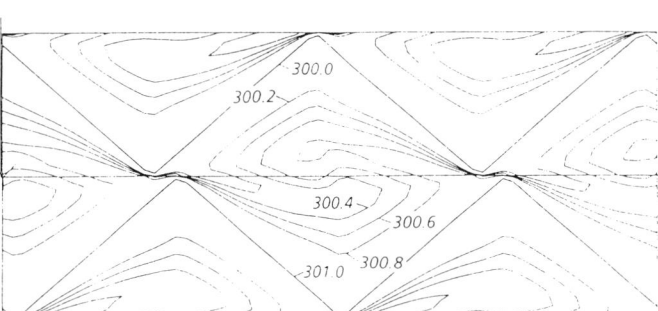

Figure 3: Temperature distribution at the outlet of the SMV mixer.

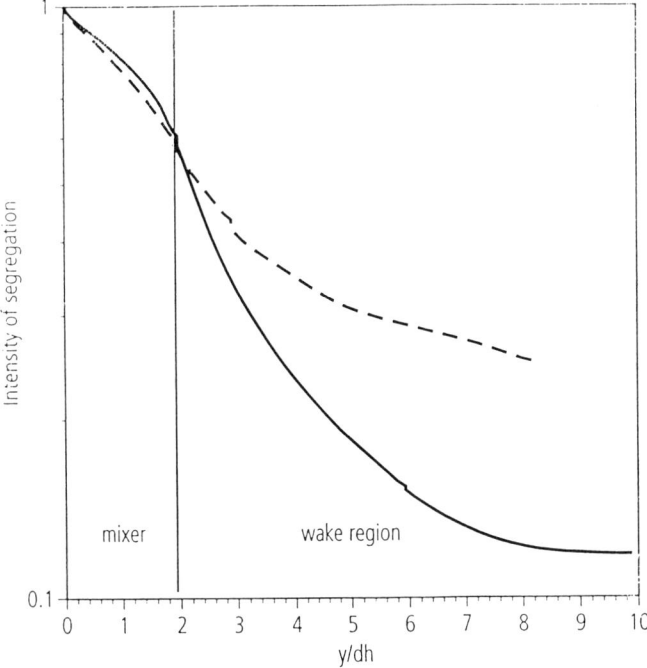

Figure 4: Computed intensity of segregation for the idealised SMV mixer without walls (   ) and for the mixer in a channel (   ).

8 Industrial Mixing Fundamentals with Applications

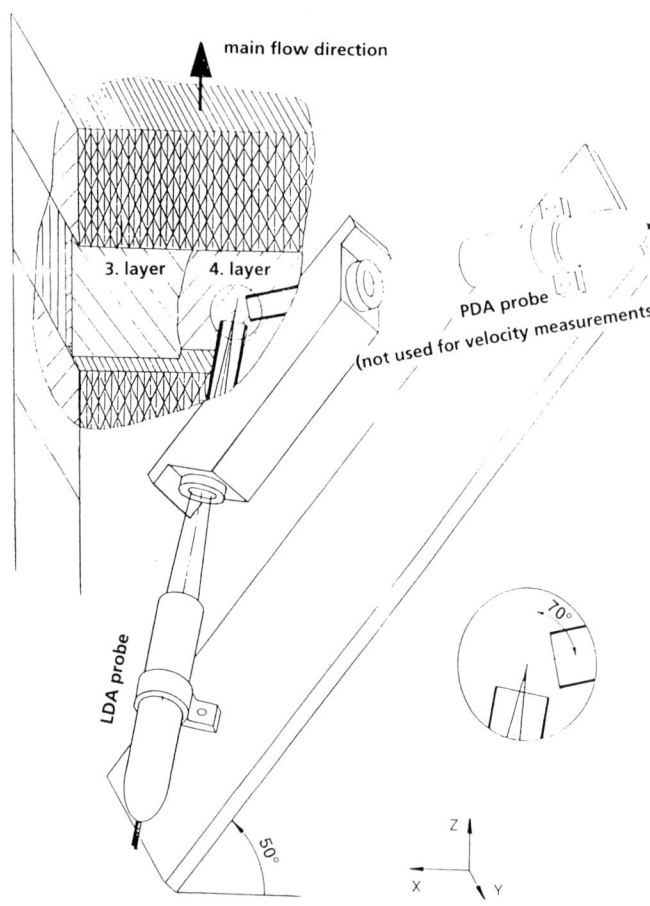

Figure 5: Experimental set-up for LDA velocity measurements in a structure similar to the SMV mixer.

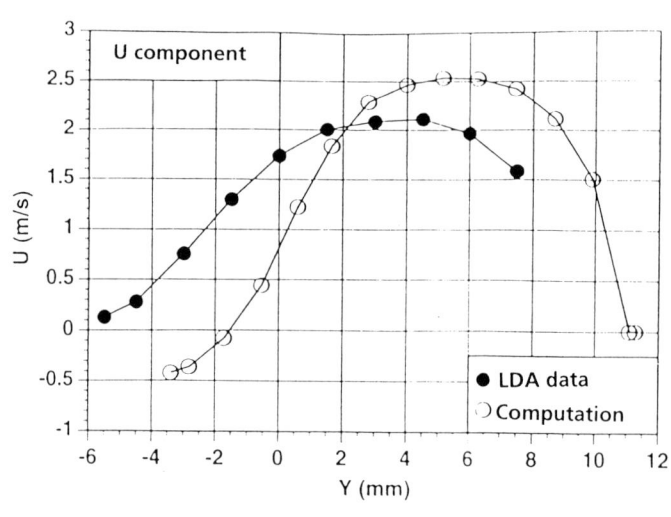

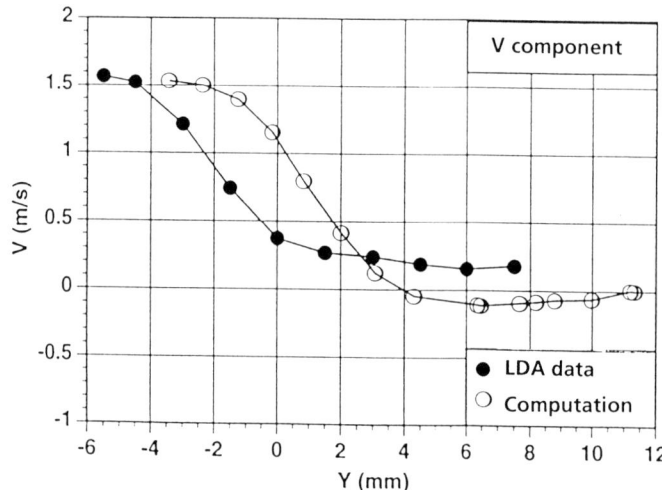

Figure 7: Comparison of measured with computed velocity components in a SMV structure.

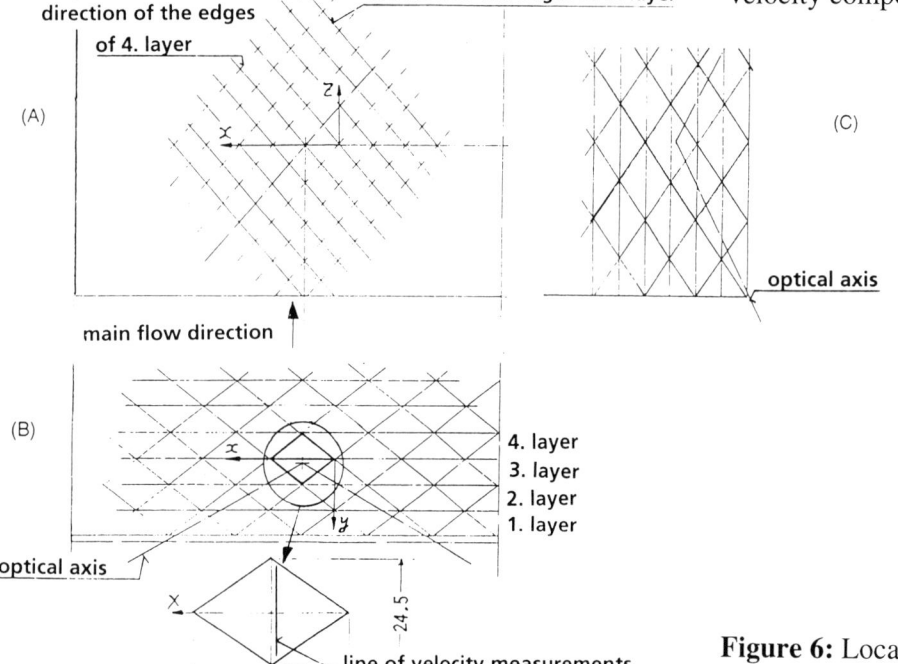

Figure 6: Location of the velocity measurement in the structure.

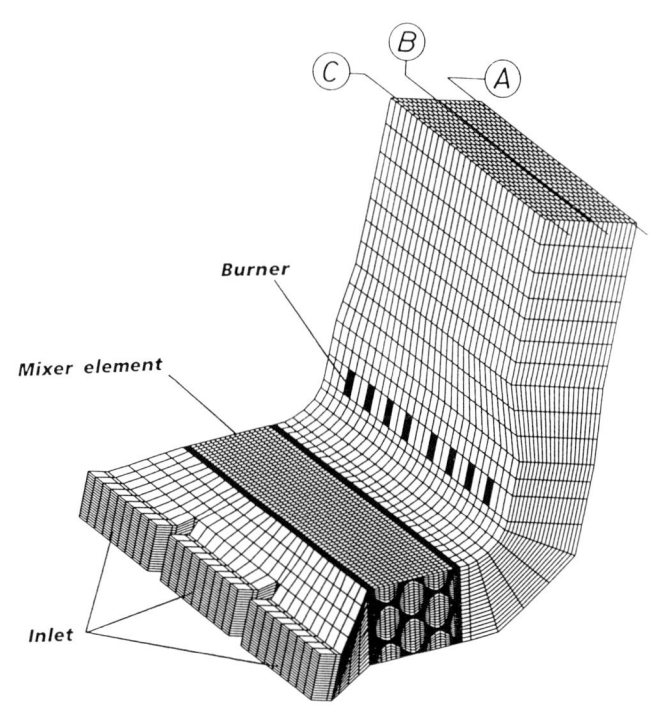

Figure 8: Grid of the inlet section of a DeNOx facility. A, B, C are the lines were data of the numerical simulation is compared to measurements.

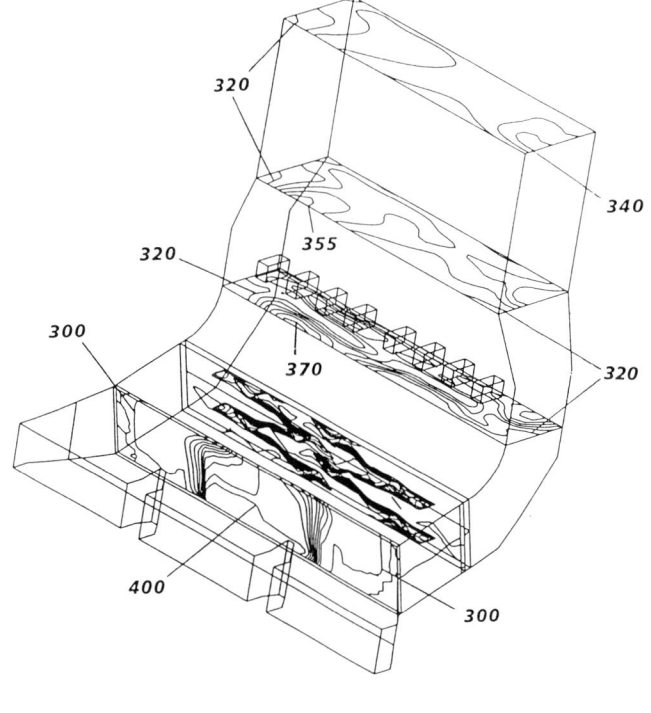

Figure 10: Temperature distributions in planes perpendicular to the main flow direction in the inlet section of a DeNOx facility with SMV mixer incorporated.

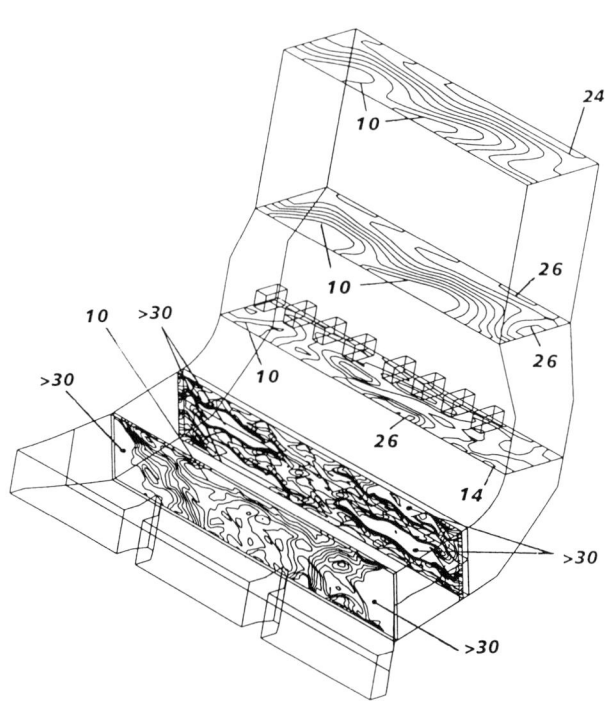

Figure 9: Velocity distributions in planes perpendicular to the main flow direction in the inlet section of a DeNOx facility with SMV mixer incorporated.

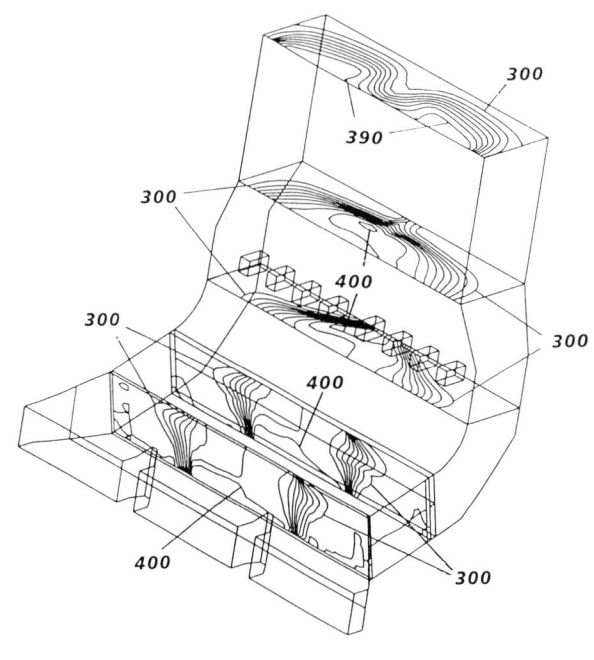

Figure 11: Temperature distributions in planes perpendicular to the main flow direction in the inlet section of a DeNOx facility without mixer.

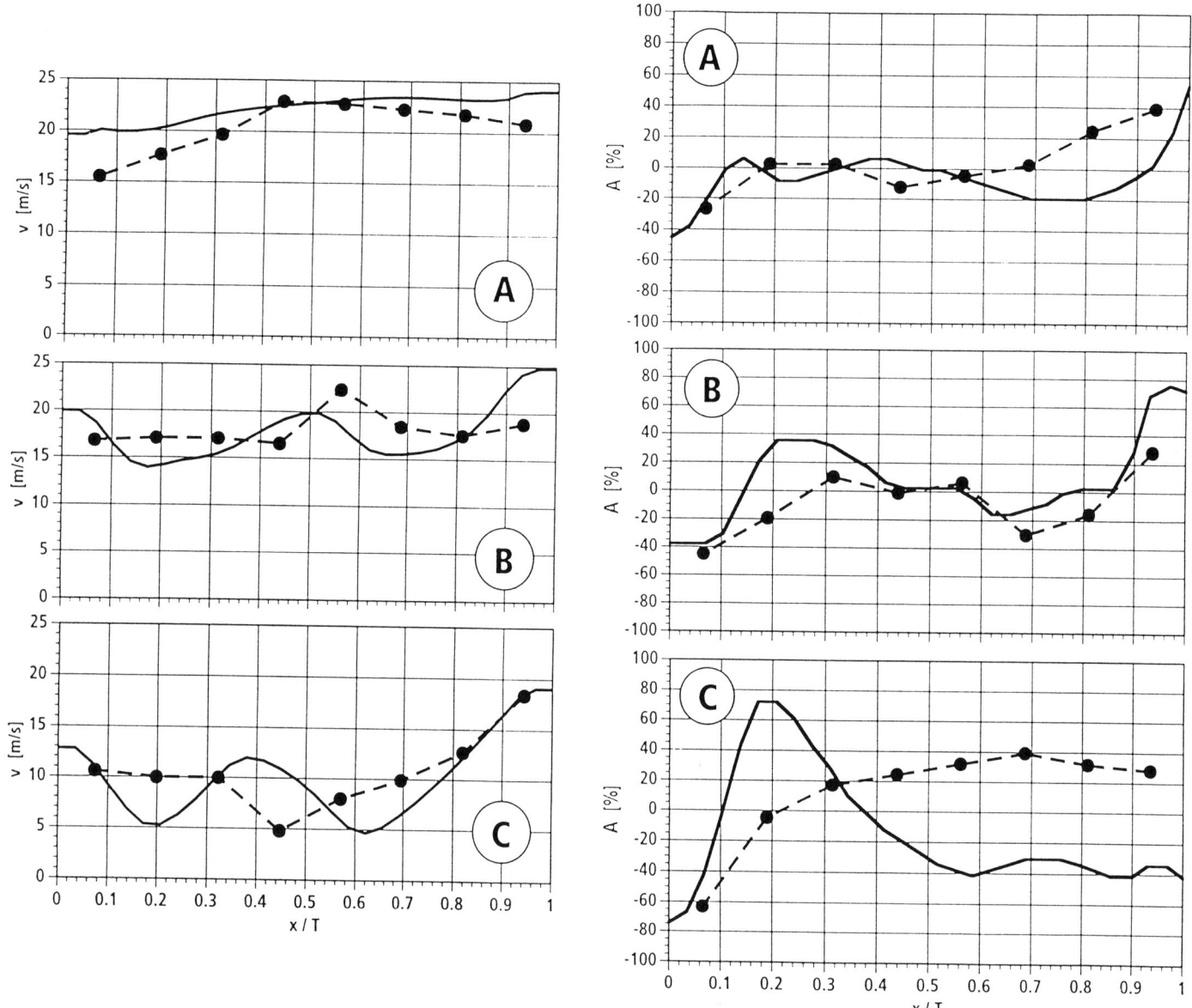

Figure 12: Left: Comparison of computed velocity to measurements along the lines A, B and C. Right: Comparison of computed temperature deviation from the mean value to measured concentration deviation from the mean value along the lines A, B and C.

Numerical Simulation and Experimental Verification of the Gas-Liquid Flow in Bubble Columns

S. Kumar[†], W.B. Vanderheyden[‡], N. Devanathan[*], N.T. Padial[‡],
M.P. Duduković[†] and B.A. Kashiwa[‡]

A numerical simulation of the two phase gas-liquid flow in a bubble column has been performed using the CFDLIB codes developed at Los Alamos National Laboratory. The simulations correspond to the flow in an air-water system operating under conditions for which experimental data is available in the literature.

The CFDLIB code for multiphase flows is based on the familiar conservation equations of the two fluid model. The interfacial momentum exchange terms account for effects of drag, lift and virtual mass effects. The code is essentially time unsteady and since the experimental information is time averaged in nature the simulated flow field is averaged over a period of several minutes for comparison purposes. Although model predictions are quantitatively in agreement with experimental data, there is a lot more to be done for incorporating improved physical models to obtain better quantitative results.

1 Introduction

The importance of the a priori prediction of the hydrodynamics of gas-liquid two phase flow in bubble columns is well recognized. A model that reliably predicts the fluid dynamics of an actual reactor is clearly advantageous in comparison to experimentation and the consequent development of a phenomenological model. Such a model calls for the numerical simulation of the complete set of governing equations of the two fluid model. Recently, there have been a number of studies in which this approach has been followed (Torvik and Svendsen, 1991, Svendsen et. al. 1992, Ranade, 1992, Lapin and Lubbert 1994, Sokolichin and Eigenberger 1994). The difficulties in handling of the interfacial exchange terms, the boundary conditions etc. make the comprehensive model a challenging problem. Consequently, there is plenty of room for research in this area. This communication reports on the results of a study in which the simulations have been made for the gas-liquid flow in bubble columns under conditions for which experimental data is available in the literature. These results constitute the first stage of a continuing program in which the complexity of the model and the physics that is accounted for is gradually being built up.

2 The CFDLIB Codes

2.1 Governing Equations

The CFDLIB codes developed at the Los Alamos National Laboratory are a versatile tool for computing three dimensional, multiphase, multispecies flows (Kashiwa et. al. 1994, Padial et. al. 1994). Although the present simulations are the results of solving only the equations of continuity and momentum, possibilities exist for including chemical reactions in the simulations and consequently to couple the kinetics to the hydrodynamics. The code is versatile in the sense that it can simulate a wide range of flow problems involving gas-liquid, gas-solid, gas-liquid-solid flows as well as packed beds. One of the characteristics of the code is that it provides a means for modeling the pressure of individual phases differently. This is

---

[†] Washington University, St. Louis, MO.
[‡] Los Alamos National Laboratory, New Mexico
[*] Amoco Oil, R & D, Naperville, IL

important for modeling the solids interaction in flows involving a solid phase. The governing equations that serve as the basis for the CFDLIB codes are:

Equation of Continuity:

$$\frac{\partial \rho_k}{\partial t} + \nabla \cdot \rho_k u_k = <\rho_o \dot{\alpha}_k> \quad (1)$$

The terms on the left hand side of the equation constitute the rate of change in mass of phase k at a given point, and the term on the right hand side is the source term due to conversion of mass from one type to the other. $\dot{\alpha}$ represents the net rate at which material k is being created.

Equation of Momentum:

$$\begin{aligned}
\frac{\partial \rho_k u_k}{\partial t} + \nabla \cdot \rho_k u_k u_k = &\text{ (rate of change in k th phase momentum)}\\
&+ <\rho_o u_o \dot{\alpha}_k> \text{ (net source of k momentum due to k mass conversion)}\\
&- \nabla \cdot <\alpha_k \rho_o u'_k u'_k> \text{ (multiphase Reynolds stress)}\\
&- \theta_k \nabla p \text{ (accln. by the equilibration pressure)}\\
&+ \rho_k g \text{ (accln. by body force)}\\
&- \nabla \theta_k (p^o_k - p) \text{ (accln. by the nonequilibrium pressure)}\\
&+ \nabla \cdot <\alpha_k \tau_o> \text{ (accln. due to average material stress)}\\
&+ <[(p_o - p)\mathbf{I} - \tau_o] \cdot \nabla \alpha_k> \text{ (momentum exchange terms)} \quad (2)
\end{aligned}$$

In the set of equations above θ_k is the expected volume fraction of phase k, while α_k is the microscopic volume fraction of phase k and the two are related by:

$$\theta_k = <\alpha_k> \quad (3)$$

where $<>$ represents an averaging process. Similarly ρ_k is the expected density of phase k and is given by

$$\rho_k = <\rho_o \alpha_k> \quad (4)$$

and u_k is the expected velocity of phase k at a given point and is obtained from:

$$u_k = \frac{<\rho_o \alpha_k u_o>}{<\rho_o \alpha_k>} \quad (5)$$

where ρ_o and u_o are the microscopic density and velocity at a given point in space.

The set of equations are exact with no approximations. Models will have to be provided for the terms with averaging operators and these models will be in terms of the expected material volume fractions θ_ks and the expected velocities u_ks.

2.2 Modeling of Interfacial Exchange and Reynolds Stress

The last term on the right hand side of the momentum equation was identified as the interfacial momentum exchange terms. Thus the effects of interfacial drag, the lift force as well as the virtual mass effects get included in this term. The drag force is expressed in terms of the phase fractions and the relative velocity between the phases as:

$$f_{kl} = \theta_k \theta_l K_{kl} (u_l - u_k) \quad (6)$$

where

$$K_{kl} = K_{lk} = \frac{\rho_c}{\theta_c} \frac{3}{8} C_D \frac{|u|}{r}$$

where C_D is the drag coefficient, and r for the case of gas-liquid flows is the bubble radius.

The modeling of the lift force is in analogy to the total force on an aerodynamic body. Consequently, the coefficient of lift is expressed as a fraction of the coefficient of drag.

The modeling of the multiphase Reynolds stress is in terms of the Prandtl's mixing length model. That is

$$- <\alpha_k \rho_o u'_k u'_k> = \rho_k l_k^2 |\nabla u_k| \epsilon_k \quad (7)$$

where l_k is a mixing length constant associated with phase k and ϵ_k is the rate of strain tensor based on the mean velocity of the phase k.

3 Numerical method

CFDLIB employs a solution method based on a class of schemes called finite-volume (FV) methods. The integral form of the conservation equations is solved on a mesh of cells each of which is treated as control volume. The variables are cell centered and the solution method utilizes an Arbitrary-Lagrangian-Eulerian split computational cycle in which the mesh is allowed to move in an arbitrary manner (Hirt et. al. 1974). This provides a means to use adaptive meshing on the flow domain. The code has a multiblock structure that enables execution on parallel computers. More details of the numerical scheme have been presented in Kashiwa et. al. (1994).

4 Simulations for the Flow in Bubble Columns

4.1 Two Dimensional Flow in a Tall Bubble Column (Chen et. al. 1989)

The first case considered is a simulation for the experiments performed by Chen et. al. (1989) in which the gas-liquid flow in a two dimensional bubble column was studied using flow visualization techniques. Chen et. al. (1989) observed the formation of von Karman vortex streets in the flow. The geometry simulated is a two dimensional bubble column with a width of 9 cm and an aspect ratio of 10. The superficial gas velocity was 0.035 m/s. The simulation considered only the effects of inertia and the effects of turbulence and added mass were neglected. The noslip boundary condition was assumed at the walls, and pressure was set to zero at the free surface. The domain was discretized into 9×90 cells. Figure 1. shows a comparison between the experimental and computed flow fields at one instant of time (approximately 90 seconds after start up) and it can be seen the code is able to predict the experimentally observed von Karman vortices by Chen et. al. (1989)

4.2 Prediction of the Data of Hills

The first case provided a qualitative comparison between the simulated results and experimental data in a planar two dimensional bubble column. The goal of the second simulation was to quantitatively compare the predicted velocity and void fraction profiles with experimental data of Hills (1974) obtained in bubble column of diameter 0.14 m with air and water as the gaseous and liquid phase respectively. The flow was assumed to be axisymmetric and consequently only one half of the column was simulated. The superficial gas velocity was 0.169 m/s and that of the liquid is zero. The initial liquid height in the column was 0.9 m. Symmetry boundary condition was assumed at the column centerline and pressure was set to zero at the free surface. Along the column walls a free slip boundary condition was used. This ignores the thin laminar layer of liquid close to the column wall. The inlet gas holdup distribution was assumed to be 0.2 and constant across the section. For this simulation the bubble size assumed was 3 mm and a constant mixing length scale of 1.5 cm was assumed. The flow domain was discretized into seven radial cells of width 1 cm and 60 axial cells of height 2 cm leading to a total of 420 cells in the domain.

The CFDLIB codes are transient in nature and therefore the simulated flow field from 10 to 20 seconds was averaged for comparison with experimental data. In Fig. 2 the time averaged liquid velocity vectors have been shown and they indicate the typical time averaged flow patterns in which the liquid ascends at the center of the column and descends at the column wall. This recirculation of the liquid phase has been experimentally documented by Devanathan et. al. (1991) using a radioactive particle tracing technique. In Fig. 3 (a) and (b) a comparison is made between the computed and experimental velocity and void fraction profiles. The computed profiles in this comparison were obtained by averaging the profiles in the middle section of the flow where the end effects are minimal. It can be seen that the predicted value for the centerline velocity is close to the experimental value. In addition the model is also able to predict the point of inversion i.e., the radial position at which the liquid velocity changes direction. The model prediction for this nondimensional radial position is approximately 0.7 which is in agreement with well known experimental results. However, there is a discrepancy in the magnitude of the liquid veloci-

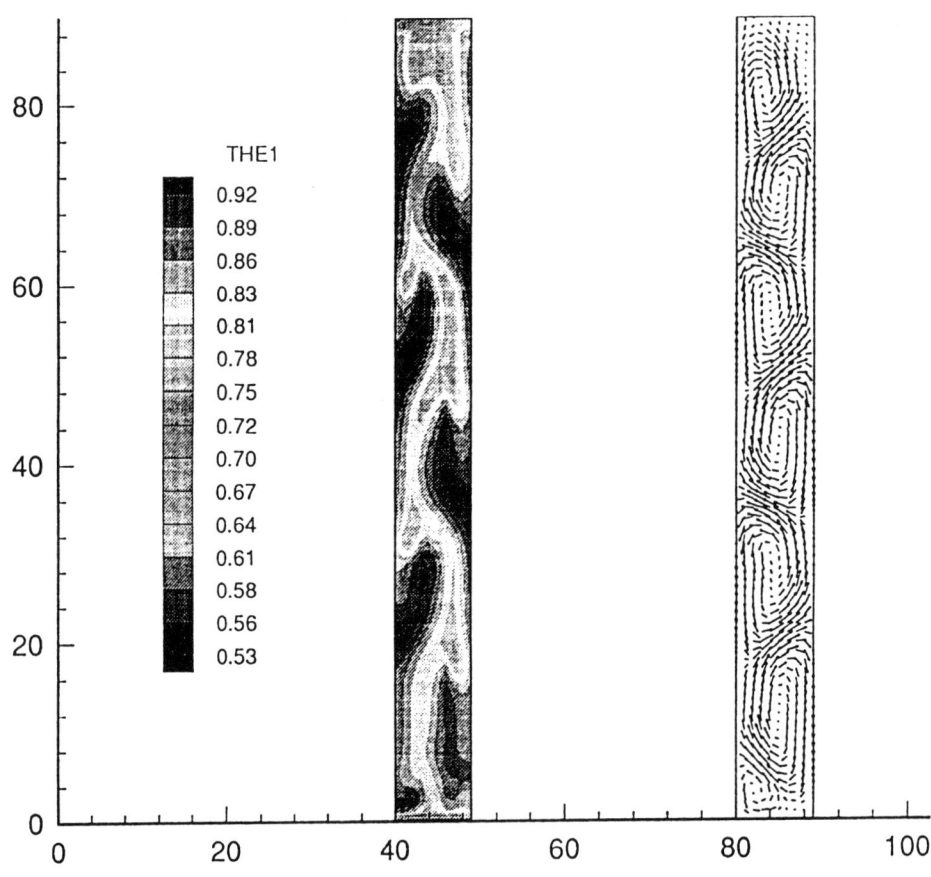

Figure 1. The simulated von Karman vortex street in a two dimensional bubble column

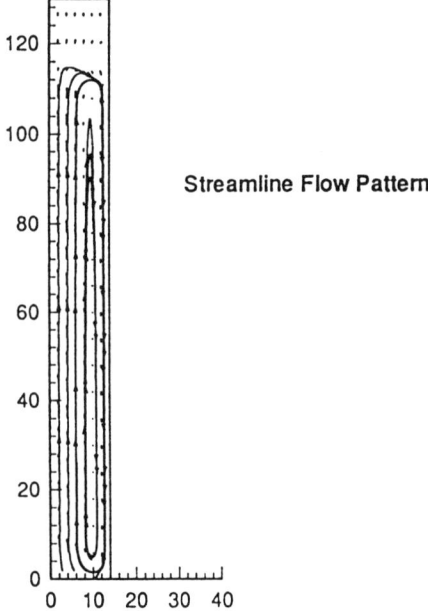

Figure 2. The stream line patterns in an axisymmetric bubble column for the experimental conditions of Hills (1974)

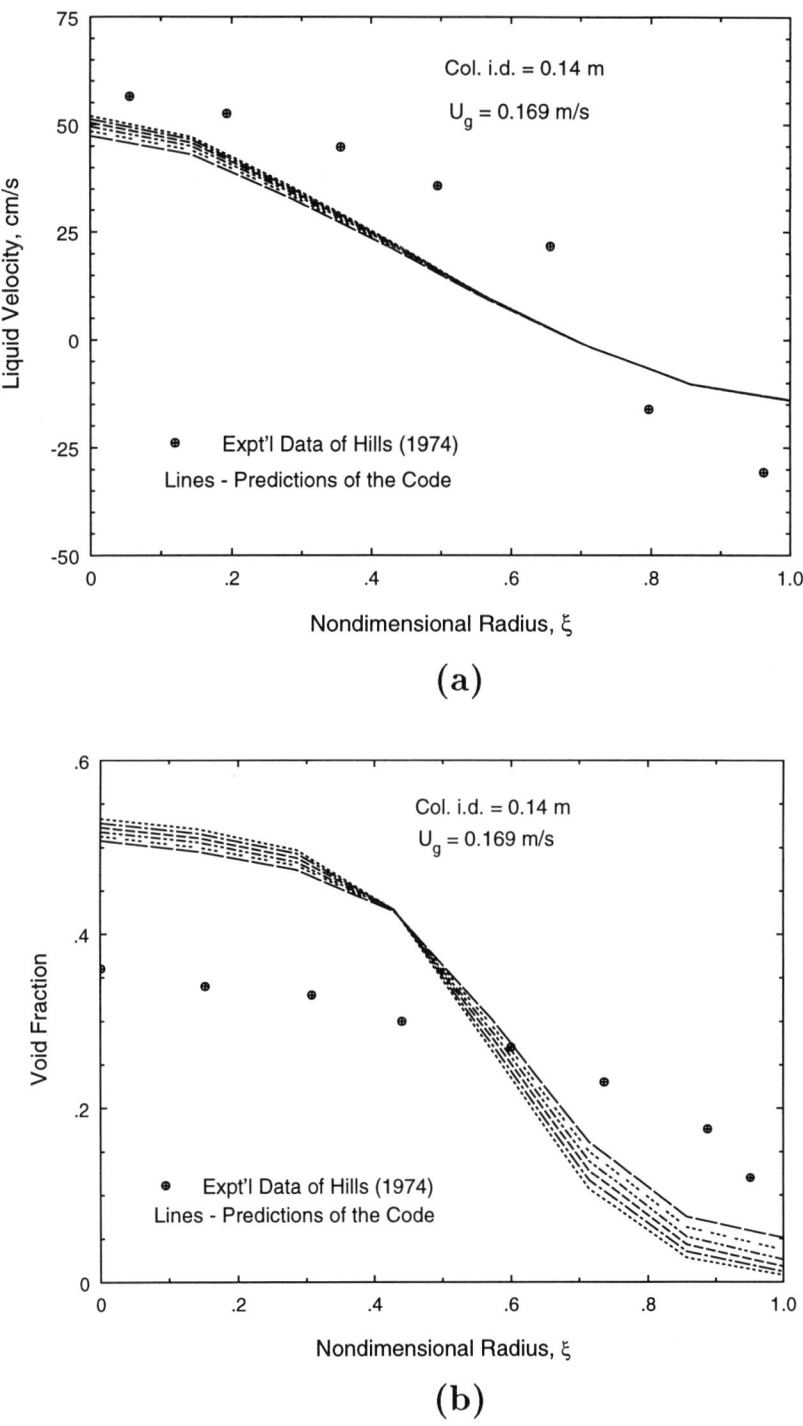

Figure 3 : Comparison of simulated and experimental velocity and void fraction profiles for the data of Hills (1974).

ties close to the wall in that they are underpredicted. The comparison for the void fraction profile indicates that the magnitude is overpredicted and the shape of radial distribution is slightly different from the expected parabolic shape. It has to be stated that this preliminary result is satisfactory in view of the simple model used for turbulence effects and the assumed wall boundary condition. However, efforts for obtaining better quantitative comparisons are underway.

4.3 Simulation of Three Dimensional Flow

The capability of the code to simulate three dimensional flows was used in the third case. The operating conditions for this experiment were as follows. The column diameter was 0.114 m and the initial liquid height was such that it gave an L/D ratio of 5.67. The superficial gas velocity for the run was 0.028 m/s. No-slip boundary conditions were used for both gas and liquid phases on the walls the vessel. Momentum exchange between the phases was modeled using a constant exchange coefficient appropriate for 2 mm bubbles in liquid. The effects of added mass were also included. No lift forces were modeled. Uniform gas inflow velocity of 2.8 cm/sec was imposed across the bottom of the column. The operating conditions were chosen to match the experimental conditions under which liquid phase velocity measurements have been by Devanathan (1991) using a Computer Automated Radioactive Particle Tracking (CARPT) facility. The simulation was run for 30 seconds of real time from a zero velocity initial condition with 20 % gas volume fraction in the liquid filled portion of the column.

A snapshot of the result at 20 seconds is shown in Fig. 4 for the liquid phase and Fig. 5 for the gas phase. In each plot, the phase velocity is represented by arrows while the volume fraction is represented by flooded contours. Some instantaneous features of the flow that can be seen in the figures is a strong down flow at the wall in the upper part of the column. Some down flow is also seen at the center of the column although it is weaker than the side downflow. Also apparent from the simulation are a number of the large scale eddies that are expected from this flow.

Detailed comparisons with experimental data will be performed after software is written to time average the computational data. This will allow more direct comparison with CARPT data.

5 Conclusions

The CFDLIB codes have been demonstrated to be a reliable vehicle for solving the conservation equations for multiphase flow systems. Some preliminary results for gas-liquid two phase flow have been obtained. The results obtained thus far are in qualitative agreement with experimental results. Improved quantitative comparisons are expected with improvements in the models for some terms for example the interfacial momentum exchange term and the multiphase Reynolds stress term. Closure modeling for these terms are ongoing, but the formalism used here is expected to serve as a guide for modeling assumptions.

6 References

Chen, J. J. J., Jamialahmadi, M. and Li, S. M., 1989, Effect of Liquid Depth on Circulation in Bubble Columns, Chem. Eng. Res. Des., Vol. 67, pp. 203-207.

Devanathan, N., 1991, Investigation of Liquid Hydrodynamics in Bubble Columns via a Computer Automated Radioactive Particle Tracking (CARPT) Facility, D.Sc. Thesis, Washington University, St. Louis, MO.

Hills, J. H., 1974, Radial Non-Uniformity of Velocity and Voidage in a Bubble Column, Trans. Inst. Chem. Engrs. Vol. 52, pp. 1-9.

Hirt, C. W., A. A. Amsden, and J. L. Cook, 1974, An Arbitrary Lagrangian-Eulerian Computing Method for all Flow Speeds, J. Comput. Phys. 14, pp. 227-253.

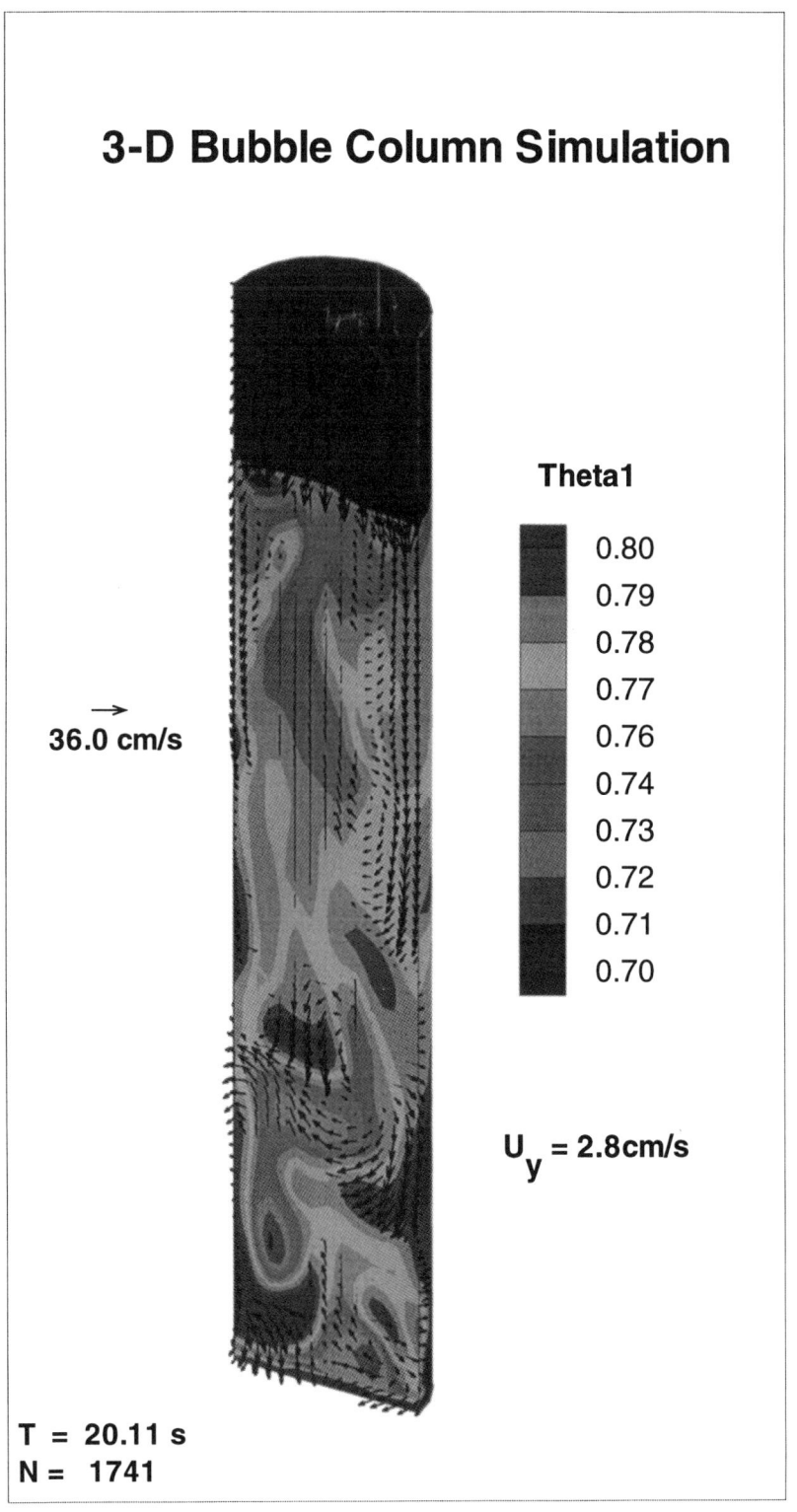

Figure 4. Snapshot of volume fraction and velocity vectors for the liquid phase

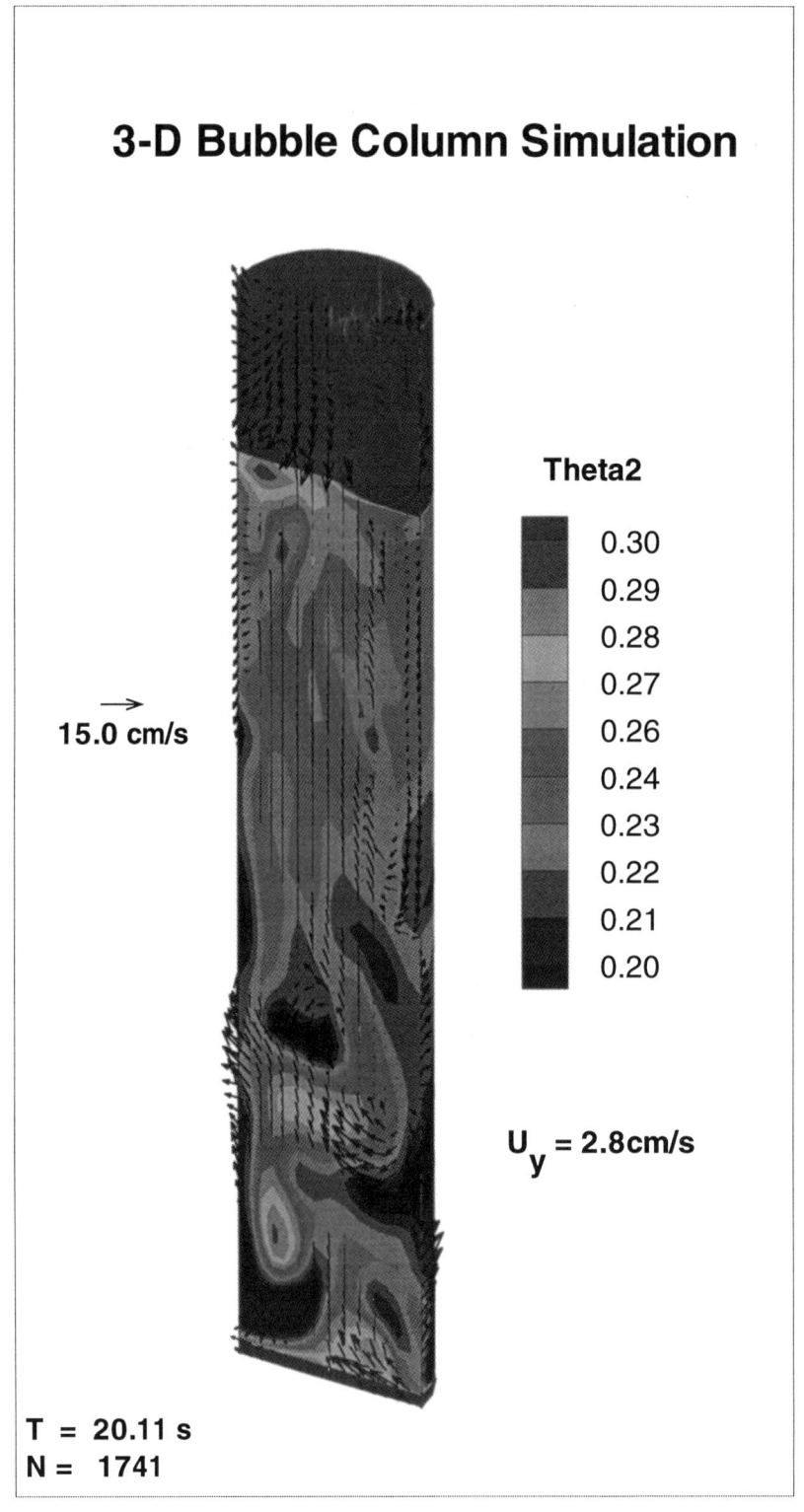

Figure 5. Snapshot of volume fraction and velocity vectors for the gas phase

Kashiwa, B. A., Padial, N. T., Rauenzahn, R. M. and W. B. VanderHeyden, 1994, A Cell-Centered ICE Method for Multiphase Flow Simulations, ASME Symposium on Numerical Methods for Multiphase Flows, Lake Tahoe, Nevada.

Lapin, A., Lubbert, A., (1994) Numerical Simulation of the Dynamics of Two-Phase Gas-Liquid Flows in Bubble Columns, Presented at ISCRE-13, Baltimore, MD, accepted by Chem. Eng. Sci.

Padial, N., B. A. Kashiwa and D. W. Kothe, 1994, , Status of CFDLIB Performance Tests on the T-3D, CRAY user group conference Tours, France.

Ranade, V. V., (1992), Flow in Bubble Columns : Some Numerical Experiments, Chem. Eng. Sci. Vol. 47, No. 8, pp. 1857-1869.

Sokolichin, A., G. Eigenberger, G. 1994, Gas-Liquid Flow in Bubble Columns and Loop Reactors, Part 1 : Detailed Modeling and Numerical Simulation, Presented at ISCRE-13, Baltimore, MD, accepted by Chem. Eng. Sci.

Svendsen, H. F., H. A. Jakobsen and R. Torvik, 1992, Local Flow Structures in Internal Loop and Bubble Column Reactors, Chem. Eng. Sci., Vol. 47, No. 9, pp. 3297-3304.

Torvik, R. and Svendsen, H. F. 1990, Modeling of Slurry Reactors. A Fundamental Approach, Chem. Eng. Sci. Vol. 45, pp. 2325-2332.

Hydrodynamic Analysis of a Two-Phase Tubular Reactor

S.L. Yarbro
Nuclear Material Technology Division, Los Alamos National Laboratory, P.O. Box 1663
Los Alamos, NM 87545

Richard Long
Chemical Engineering Department, New Mexico State University, P.O. Box 30001 Dept. 3805,
Las Cruces, NM 88003

Many typical chemical engineering operations are multiphase tubular reactors, such as packed columns, pipes and others. An important parameter is interfacial area, which determines the rate of heat and mass transfer. In many cases, the models for determining area are empirical and can only describe the cases for which there is experimental data. In an effort to understand multiphase reactors and the mixing process better, a model has been developed as part of a research effort to extend the current empirical correlations for interfacial area in a in-line static mixer. For this work, a simple hydrodynamic model has been developed and tested for a Kenics in-line mixer. The model equations are presented along with comparisons to existing and new experimental data.

INTRODUCTION

In the chemical process industry, it has been estimated that $ 10 billion dollars are wasted each year because of inefficient mixing. This is likely due to a lack of process knowledge that results in an over-design to compensate or under-design that produces recycle or high reagent losses. Most design procedures are based on either empirical correlations or ideal models which ignore most of the non-idealities that are involved with actual equipment. Therefore, examining methods of extending these models to include actual system effects is economically justifiable.

For design involving multiphase systems, interfacial area is a key parameter for calculating mass, heat and momentum transfer rates. This study focuses on a two-phase system of immiscible fluids, such as those typically used for liquid-liquid extraction. For these systems, empirical design correlations are normally developed for specific equipment such as packed columns or mixer-settlers [1]. This restricts the design to systems that are within the experimental data. Consequently, there is a need for a more general approach.

BRIEF REVIEW OF RECENT WORK

Several groups have successfully studied the problem of empirically predicting interfacial area in systems of immiscible fluids. We are interested in systems where the contactor is a "tubular reactor" such as columns, packed beds or in-line mixers. In-line mixers, such as Kenics or Sulzer mixers, are attractive because they are simple, have low residence times, and are easy to operate and maintain.

Middleman [2] measured drop size distributions for several systems with viscosity's ranging from 0.6 to 26 cp and interfacial tensions from 5 to 46 dynes/cm in a Kenics mixer. He correlated his data with the following equation:

$$\frac{D_{32}}{D_0} = 0.49 \text{We}^{-3/5} \quad (1)$$

Middleman derived this equation by using the Kolmogorov theory. Middleman assumed that the dispersive energy causing drop breakage was due to the inertial subrange eddies and that the drop was stabilized only by the interfacial tension. The above correlation begins to fail as the drop viscosity increases, although it is accurate for systems of relatively inviscous fluids. His work also showed that drop size distribution was independent of dispersed phase fractions up to 25%.

Berkman and Calabrese [3] extended these correlations by examining the effects of viscosity and hydrodynamic conditions on drop size distributions produced by a Kenics mixer. He found a slight dependence on hydrodynamic conditions for Re > 12,000 and a measurable dependence on viscosity with viscosity's > 20 cp. They correlated their data to the following

$$\frac{D_{32}}{D_0} = 0.49 \text{We}^{-3/5} \left[1 + 1.38 \text{Vi} \left(\frac{D_{32}}{D_0}\right)^{1/3}\right]^{3/5} \quad (2)$$

They concluded that residence time and energy dissipation rate were important to the drop production in the contactor.

Al Taweel and Walker [4] measured drop size distributions for dilute mixtures in a Lightnin "In-Line" mixer. They were able to correlate their data in terms of an energy dissipation rate or Weber number. They discovered that the assumption of equilibrium for two-phase mixing systems is not always correct. Long and Reimus [5] and Long, et.al. [6] observed that equations (1) and (2) do not fully account for the effects noted at a tee junction. In particular, they observed that equations (1) and (2) must be corrected by additive terms that were proportional to dispersed phase volume fraction. They attributed this effect to the geometry of their system, where one would expect the breakage time to be large compared to the residence time. This is an important effect in such systems because it implies that even when the Weber number increases without bound, it is impossible to drive the droplet size to zero. Rather an additive term reflecting the lower limit must be included. Al Taweel and Walker did not discover this effect because all of their work was done at a constant and low dispersed phase fraction of 1%. The experimental data is compiled in Figure 1.

EXPERIMENTAL EQUIPMENT

In this study, a 12-element, 0.635 cm diameter, 15.24 cm long Kenics mixer with a pitch ratio of 0.8 was used. Micropump gear metering pumps (2500 ml/min capacity) were used with stainless steel pump heads. All hardware was Teflon tube with compression fittings. Five-liter glass carboys were used to hold the phases. Flowrates and phase fractions were measured by timing the flow into a graduated cylinder and measuring the resultant volumes. The drop size distributions were measured using a Kodak COHU electronic camera with a CCD speed of 30 frames-per-second, a electronic shutter speed of 1/10,000 of a sec and a drop resolution of approximately 30 microns. The light source was a 100 W incandescent bulb. The bulb was placed behind a shield with a slit width of 20 mm and a light diffuser of either frosted glass or paper. The light source was then oriented to produce some contrast in the image to enhance the drops. The data was recorded on VCR tape and processed on a Gateway 2000 PC installed with a Raptor frame capture board and Image Pro software for image manipulation. Drop sizes were counted both manually and electronically to ensure repeatability. Reagent grade dodecane and distilled water were the fluids used in this study. The equipment is shown in Figure 2.

MATHEMATICAL MODEL

The general balance equations for a multiphase system are shown below [7]. For mass

$$\frac{\partial \rho_j}{\partial t} + \overline{\nabla} \cdot (\rho_j \phi_j \overline{v}_j) = 0 \quad (3)$$

with ϕ_j being the phase fraction and the momentum balance

$$\frac{\partial(\rho_j \phi_j \overline{v}_j)}{\partial t} + [\overline{\nabla} \cdot \rho_j \phi_j \overline{v}_j \overline{v}_j] = -\phi_j \overline{\nabla} p - \phi_j [\overline{\nabla} \cdot \overline{\overline{\tau}}] + \phi_j \rho_j \overline{g} + M_j \quad (4)$$

For the purpose of this model, it was assumed that the system was at steady-state, the fluids were incompressible and gravity effects could be neglected. Because there is no phase change, the pressure difference between the fluids is negligible and not included.

To develop the constitutive equations for shear, it was assumed that the flow was turbulent and therefore simple friction factor correlations could be used to predict wall and interfacial shear. Specifically for the interfacial shear, an approach used for submerged objects [8] was used. In the turbulent limit near the entrance of the reactor, the friction factor becomes equal to approximately 0.44. The constitutive equations become

$$\tau_{c_w} = \frac{1}{2} v_c^2 f \rho_c \quad (5)$$

$$\tau_{d_w} = \frac{1}{2} v_d^2 f \rho_d \quad (6)$$

$$\tau_i = 0.22 \rho_c (v_c - v_d)^2 \quad (7)$$

Other forces such as the Basset and lift forces are assumed to be small with regard to the interfacial shear especially for small phase fractions. The viscous shear, τ_{zz}, which comes from expansion of the Div τ term, is also assumed to be small for inviscid fluids.

Drop coalescence is assumed to be negligible for small phase fractions and only drop breakage is considered. Based on previous work, drop breakage appears to be a function of the energy dissipated by the continuous phase and the contact time. Because of the large velocity difference between the two fluids at the entrance of the reactor, interfacial shear along with wall shear, would be an important drop production mechanism. With this assumption and the assumption that the drop size is a function of drop surface energy $(4\pi D^2 \rho \sigma)$ and surface turbulent forces $(4/3\pi D^3 \rho \varepsilon)$ the following relationship for drop size can be developed.

$$\frac{D_{32}}{D_0} = \frac{3\sigma}{\rho_c \tau_{res} \varepsilon} \quad (8)$$

$$\varepsilon = \frac{Q \int_0^z \left(\tau_i + \tau_{d_w} \right) dz}{\rho_d A L} \quad (9)$$

$$\tau_{res} = \frac{Volume}{Flowrate} \quad (10)$$

With these simplifications, the following one-dimensional model can be derived

$$\rho_j \left[v_j \frac{\partial \phi_j}{\partial z} + \phi_j \frac{\partial v_j}{\partial z} \right] = 0 \quad (11)$$

$$\rho_j \left[v_j^2 \frac{\partial \phi_j}{\partial z} + 2\phi_j v_j \frac{\partial v_j}{\partial z} \right] =$$

$$-\phi_j \frac{\partial p}{\partial z} + \frac{1}{R} \tau_i - \frac{1}{R} \tau_{j_w} \quad (12)$$

Therefore, equations eight through 12 make up the model along with the following boundary conditions

$$\phi_j(0) = 1.0 \quad (13)$$

$$v_j(0) = v_{j_0} \quad (14)$$

RESULTS AND DISCUSSION

Drop size distributions were measured for the dodecane-water system at two different velocities and a phase fraction of 0.067. The results are compared with data from Middleman, and Berkman and Calabrese below. The physical properties of the dispersed phases are compiled in Table 1. The continuous phase in all the studies was water.

The data is reasonably correlated by the Weber No. alone. Therefore, the assumption that the viscous terms are small compared with the inertial terms is reasonable. The model equations were solved using the numerical integration routine in Mathematica 2.2 for the Macintosh.

The velocity profiles are shown in Figures 3 and 4. The two phase velocities can accelerate because the phase fraction is allowed to vary. As shown, the momentum of the dispersed phase rapidly increases and it accelerates. The continuous phase shows a corresponding deceleration, but this is small compared to its overall momentum. The rapid acceleration of the dispersed phase then causes an extremely high interfacial shear which causes most of the droplet production early in the reactor. As the velocity of the two phases becomes the same, the dominant drop production mechanism becomes the wall shear.

The interfacial shear and wall shear are integrated over the length of the reactor to get the total energy dissipated by the continuous phase which contributes to drop production. This value is then corrected by the residence time. The results of the drop size model are shown in Figure 5.

The model results compare favorably with Yarbro and Long's experimental results but are high for Berkman and Calabrese and Middleman. Interfacial shear is important but does not appear to be a dominant drop production mechanism except at the entrance of the reactor. Wall shear, particularly for higher flowrates, has as much or more of a contribution to droplet production. Assuming viscous forces are small compared to shear forces appears to be reasonable for the conditions of this study. However, the results indicate that viscous effects should be included in a more general model. This approach has the advantage that it does not have any empirical constants except for conventional friction factors. It can also be extended to predict drop size as a function of reactor length to allow accurate mass and heat transfer calculations. Future work will include accounting for viscous effects and more accurate accounting for turbulent fluctuations of energy dissipation.

ACKNOWLEDGEMENT

We would like to thank Mr. John Macdonald for his help with assembling and configuring the computer and camera system for the experimental work completed in this study.

NOTATION

A = Cross-sectional area

D_{32} = Sauter mean diameter,

$$\sum_{k=1}^{m} n_i D_i^3 \Big/ \sum_{k=1}^{m} n_i D_i^2$$

D_0 = Reactor diameter

L = Reactor length

f = Friction factor

Q = Volumetric flowrate

Vi = Viscosity ratio

v_j = Velocity of the jth phase

N_{We} = Weber no.

τ_{jw} = Wall shear of the jth phase

τ_i = Interfacial shear

σ = Surface tension

ρ_j = Density of the jth phase

ε = Energy dissipation

ϕ_j = Phase fraction of the jth phase

LITERATURE CITED

1. Thornton, J.D., *Nucl.Eng.*, July, 156, (1956)

2. Middleman, S., *Ind.Eng.Chem.Process Des.Dev.*, **19**, 91, (1974)

3. Berkman, P.D. and R.V. Calabrese, *AIChE J.*, **34**, 602, (1988)

4. Al Taweel, A.M. and L.D. Walker, *Can.J.Chem.Eng.*, **61**, 527, (1983)

5. Long, R.L. and P. Reimus, *Chem.Eng.Comm.*, **3**, 1, (1992)

6. Long, R., Tavarez, I., Lin, W. and P. Reimus, "Interfacial Area Production in Tees and Jets" in <u>Process Mixing: Chemical and Biochemical Applications</u>, *AIChE Symp.Ser.*, **88**, No. 286, 65, (1992)

7. Lahey, R.T., *AIChE J.*, **37**, 123, (1992)

8. Bird, Stewart and Lightfoot, <u>Transport Phenomena</u>, Chap. 6, (1960)

Table 1

Physical Properties of Selected Dispersed Phases

Study	Material	Density (g/cm^3)	Viscosity (cp)
Yarbro and Long	Dodecane	0.748	1.35
Berkman and Calabrese	p-Xylene	0.856	0.64
Middleman	Cyclohexane	0.760	0.66

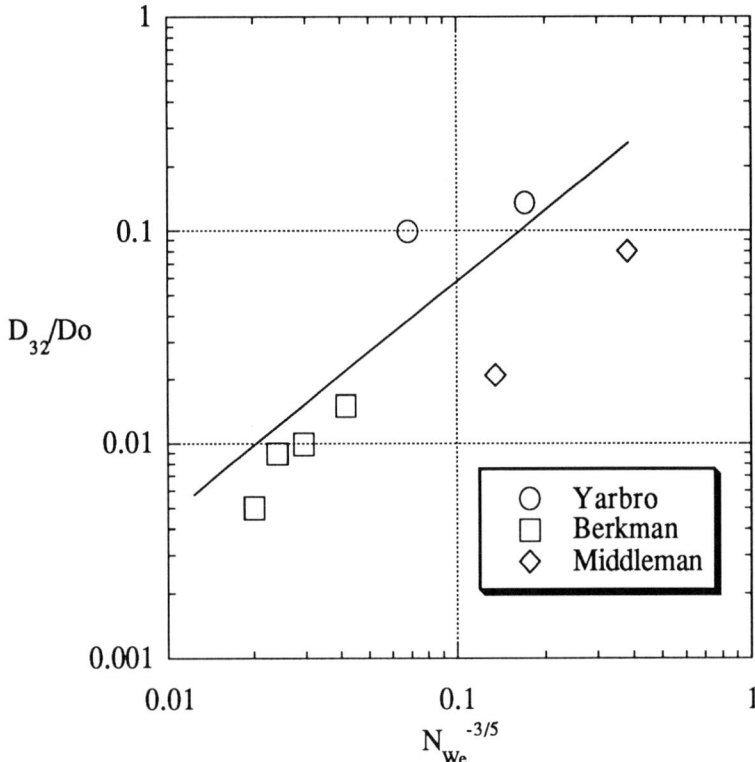

Figure 1. Comparison of Several Drop Size Distributions as a Function of N_{We}

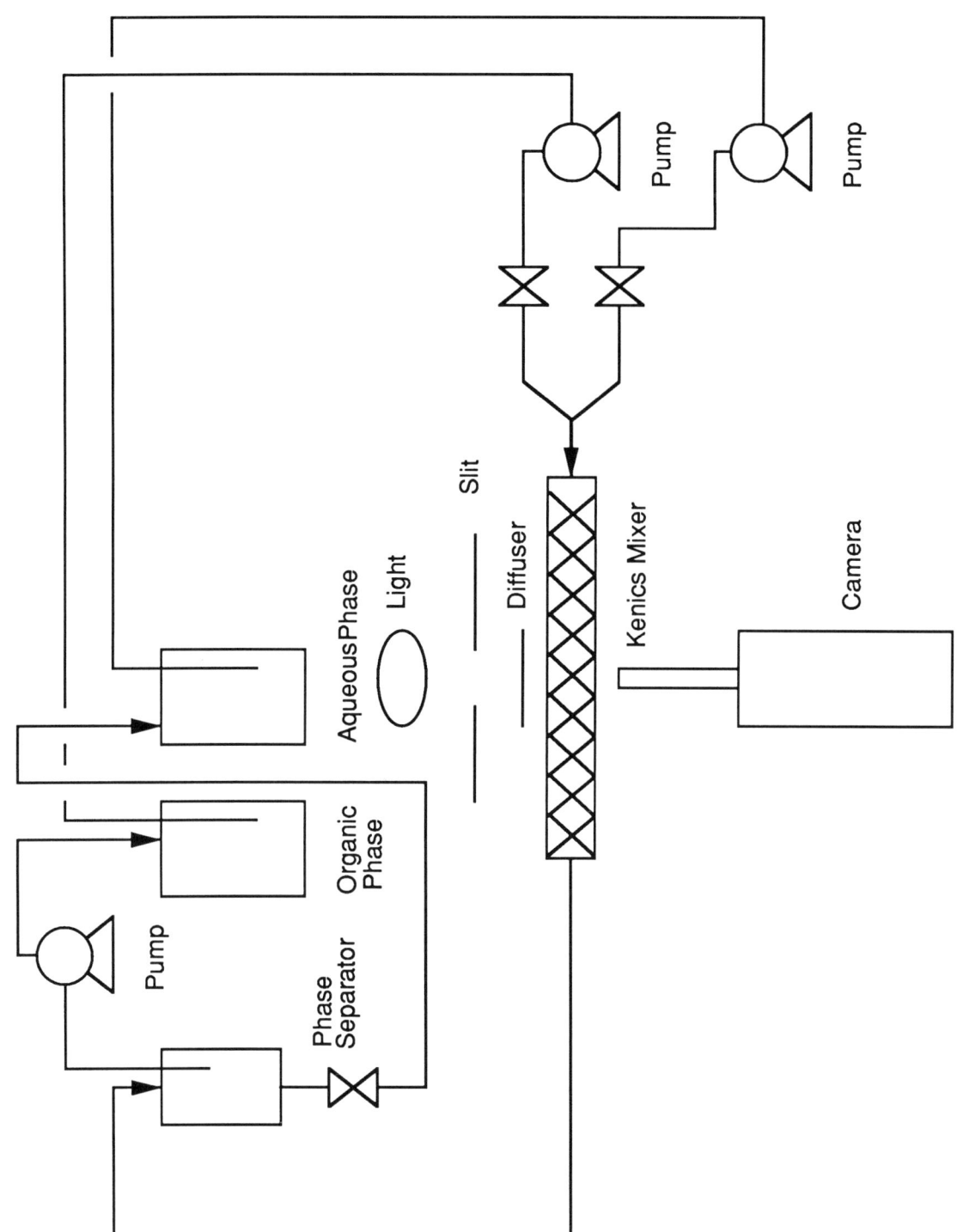

Figure 2. Schematic of the Experimental Equipment

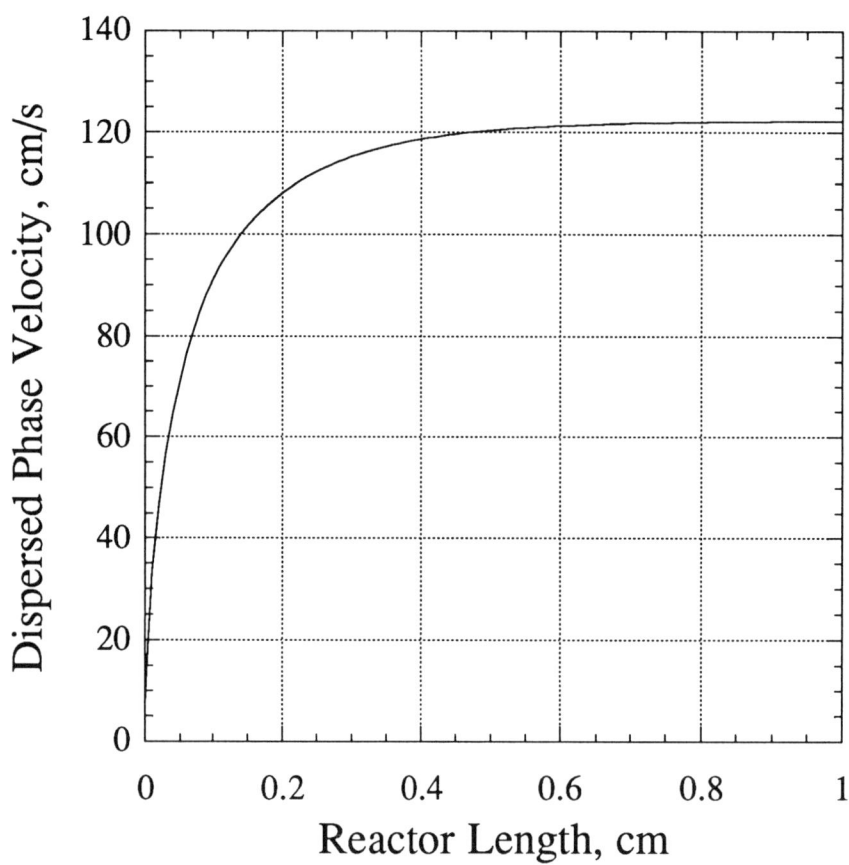

Figure 3. Acceleration of the Dispersed Phase $N_{We} = 165$

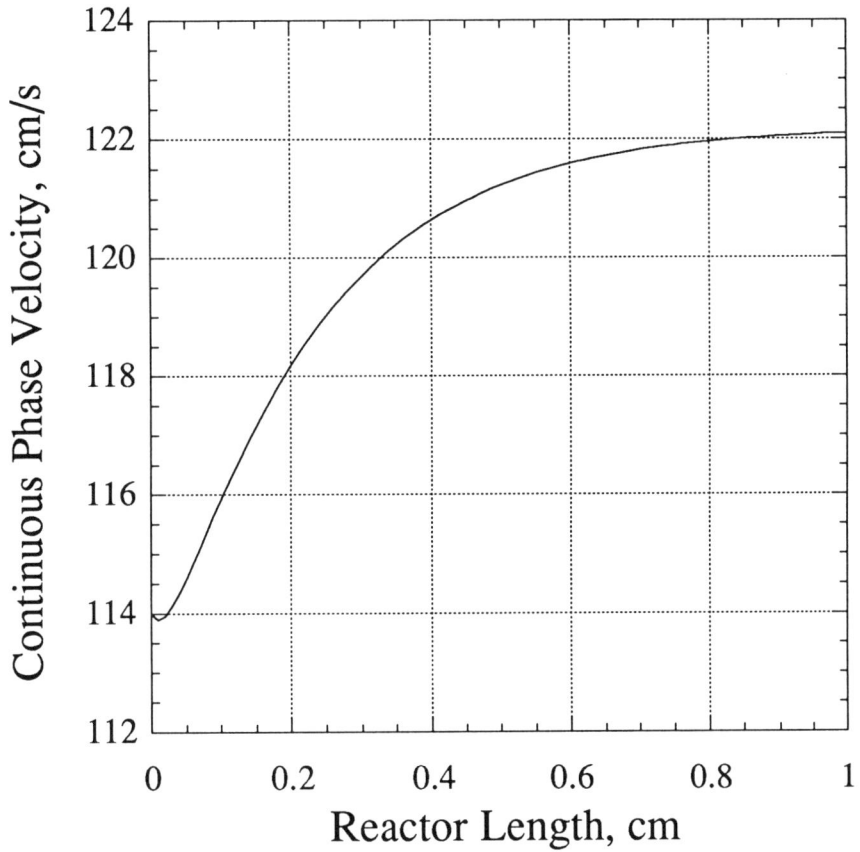

Figure 4. Corresponding Deceleration of the Continuous Phase
$N_{We} = 165$

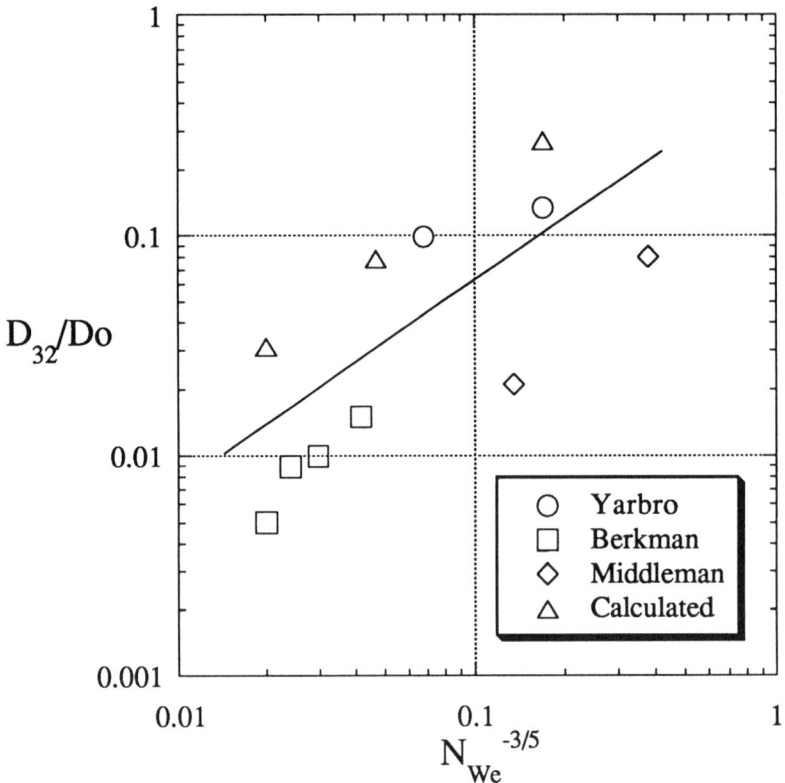

Figure 5. Comparison of Experimental and Calculated Drop Diameters

PDF Modeling of Turbulent Mixing and Chemical Reactions in a Tubular Jet Reactor

Kuochen Tsai and Rodney O. Fox*
College of Engineering, Kansas State University, Manhattan, KS 66506

One advantage of probability density function (PDF) methods over traditional moment methods is their ability to cope with chemical reaction terms without modeling. Therefore, they are an ideal tool for the study of different assumptions in turbulent mixing models. In this paper, a Lagrangian velocity-composition PDF (LVCPDF) code and a Lagrangian composition PDF (LCPDF) code are used to simulate a series-parallel reaction in a single-jet tubular reactor and the results are compared with experimental data. Both methods give very similar results despite the different closures adopted. By comparing the results with the LVCPDF code, the gradient-diffusion assumption in the LCPDF formulation is verified for incompressible flows. The turbulent flow field is obtained from the k-ε model using FLUENT and the chemical reaction terms are efficiently dealt with using a look-up table.

INTRODUCTION

The simulation of turbulent reacting flows is difficult because of the randomness of the flow field and the complex chemical processes involved. Traditionally these problems are solved using the moment method which deals with averaged quantities of the flow field by forming a set of deterministic equations from the Navier-Stokes equation. Major difficulties encountered in this formulation are the nonlinear turbulent advection, micromixing and chemical reaction terms.

One great advantage of probability density function (PDF) methods over moment methods is their ability to cope with the nonlinear reaction terms without modeling (Dopazo, 1973). In the PDF approach, a joint PDF is used to describe the variables of a turbulent flow field, thus it includes information about all the moments. However, this approach has a closure problem similar to the BBGKY hierarchy in statistical mechanics (Lundgren, 1969; Monin and Yaglom, 1967). For example, when using a one-point statistical description, knowledge about its neighboring points (or its gradient) is necessary to describe the diffusion process.

Three processes are involved in a turbulent reacting flow, namely, convection, diffusion and reaction. By using the PDF approach, the reaction terms are always closed, and with the inclusion of velocity as phase variables, the turbulent convection terms are also closed. For this reason, a state-of-the-art joint Lagrangian velocity-composition PDF (LVCPDF) code (Pope, 1991) was adopted to study the effect of scalar mixing on chemical reactions in an earlier work (Tsai and Fox, 1994). Although this formulation leaves unclosed terms in velocity space, they can be modeled using well-established turbulence models, such as the $k-\epsilon$ model (Pope, 1985).

The LVCPDF code is highly effective for studying models for scalar mixing rate; however, it is prohibitively expensive for 3-dimensional simulations. In addition, because of the finite sampling size in each finite-difference cell, the coupling of scalar and velocity variables cannot be completely separated in phase space. In other words, even though theoretically it is possible to close the turbulent advection terms in a LVCPDF formulation, in practice, such a task is difficult to carry out for inhomogeneous flows. In order to further reduce the computational expense, a joint Lagrangian composition PDF (LCPDF) code, which is derived by integrating the velocity-composition PDF equation over velocity space, is tested in this work. Using this formulation, velocity variables are removed and the gradient-diffusion assumption is introduced to account for the cross-correlations between composition and velocity. In order to describe the scalar fields in a continuous manner, a Lagrangian description is adopted. Unlike Eulerian composition PDF codes, this formulation allows a continuous distribution of stochastical particles in the simulation domain, thereby, yielding a better description of the scalar field. The applicability of the gradient-diffusion assumption can be evaluated through comparison with the results

---
*Author to whom all correspondence should be addressed.

from the LVCPDF code. In this study, both methods are compared with the experimental data of Li and Toor (1986).

The series-parallel chemical reaction in Li and Toor's experiment has the following form:

$$A + B \xrightarrow{k_1} R,$$
$$B + R \xrightarrow{k_2} S. \qquad (1)$$

Because $k_1 \to \infty$ and $k_1 \gg k_2$, the production of S is extremely sensitive to scalar mixing. This property makes this reaction a good test case for mixing rate models. With the experimental data and the simulation results of the LVCPDF code, a detailed comparison can be made to verify the gradient-diffusion assumption in the LCPDF formulation. The chemical reaction terms in both PDF codes were evaluated by interpolating a chemical look-up table (Tsai and Fox, 1994). Due to the fact that chemical reactants have finite domain (e.g. between 0 and 1), the changes of reactant concentrations in a small time step for all possible values within the finite domain can be stored in the table.

In this study, our intention is to examine the two PDF codes when applied to chemical reactions in a tubular reactor. The model for scalar mixing rate was chosen to be $C_\phi \sqrt{k}/d$ which exhibits the best fit with the experimental data (within 2% accuracy) when compared with other models (Tsai and Fox, 1994). The flow field of Li and Toor's experiment (1986) were simulated using FLUENT with the standard $k - \epsilon$ model (Launder and Spalding, 1974).

PROBLEM FORMULATION

The governing equations for the transport of reactive scalars in a turbulent flow with constant density and isothermal chemical reactions can be written for Eq. 1 as

$$\frac{\partial u_i}{\partial t} + u_j \frac{\partial u_i}{\partial x_j} = -\frac{\nabla p}{\rho} + \frac{1}{\rho} \frac{\partial \tau_{ij}}{\partial x_j} + g_i,$$
$$i = 1, 2, 3, \qquad (2)$$

$$\frac{\partial \phi_k}{\partial t} + u_j \frac{\partial \phi_k}{\partial x_j} = D_{\phi_k} \nabla^2 \phi_k + S_k(\phi),$$
$$k = A, B, R, S, \qquad (3)$$

$$\nabla \cdot \mathbf{u} = 0, \qquad (4)$$

where τ_{ij} is the viscous stress tensor defined by

$$\tau_{ij} = \mu \left(\frac{\partial u_i}{\partial x_j} + \frac{\partial u_j}{\partial x_i} \right) - \frac{2}{3} \mu \frac{\partial u_l}{\partial x_l}, \qquad (5)$$

u_i are the velocity components, p is the pressure, ρ is the fluid density, ϕ_k represents the chemical species carried by the fluid, D_{ϕ_k} is the diffusivity of ϕ_k and $S_k(\phi)$ are the chemical production terms which can be written as

$$\begin{aligned} S_A &= -k_1 \phi_A \phi_B, \\ S_B &= -k_1 \phi_A \phi_B - k_2 \phi_B \phi_R, \\ S_P &= +k_1 \phi_A \phi_B - k_2 \phi_B \phi_R, \\ S_S &= +k_2 \phi_B \phi_R, \end{aligned} \qquad (6)$$

where $k_1 = 5.0 \times 10^6$ m^3/kmol$\cdot$s and $k_2 = 1.8 \times 10^3$ m^3/kmol$\cdot$s (Li and Toor, 1986). The reactive scalars considered here are passive so there is no interaction between the scalars and the velocity field. The chemical reaction is isothermal, which excludes the consideration of any thermodynamic property changes. In Li and Toor's experiment, the reactions occur in liquid phase and the temperature change in the reactor is negligible; therefore, the passive scalar assumption is reasonable.

PDF METHODS

The PDF simulations were performed using a set of notional Lagrangian stochastic particles moving in the computational domain. Each of the particles obeys a set of stochastic differential equations (SDEs) which mimic the PDF transport in the physical and the phase space (Pope, 1985). Before the SDEs are formulated, the PDF balance equation can be derived from the Navier-Stokes equation and the conservation equations for passive scalars (Pope, 1985). The governing equations of the velocity-composition and the composition PDF are formulated as follows.

Velocity-Composition PDF Equation

Detailed descriptions of PDF methods can be found in O'Brien (1980) and Pope (1985). Here, we only outline the formulation used in the present study. Following O'Brien (1980), in order to derive the evolution equation for the velocity-composition PDF a fine-grained density is defined,

$$\mathcal{F} = \delta(\mathbf{u}(\mathbf{x}, t) - \mathbf{v}) \delta(\phi(\mathbf{x}, t) - \psi), \qquad (7)$$

where $\mathbf{u}$ and ϕ are physical variables, and $\mathbf{v}$ and ψ are the corresponding variables in phase space. This function could be thought of as a signal which records the events occurring in physical space into phase space. The PDF can be derived by taking the ensemble average of this function:

$$f(\mathbf{v}, \psi) = \langle \mathcal{F} \rangle. \qquad (8)$$

The PDF balance equation can be derived by first taking the time derivative of $\mathcal{F}$,

$$\frac{\partial \mathcal{F}}{\partial t} = \delta(\boldsymbol{\phi}-\boldsymbol{\psi})\frac{\partial \delta(\mathbf{u}-\mathbf{v})}{\partial \mathbf{u}}\frac{\partial \mathbf{u}}{\partial t} + \delta(\mathbf{u}-\mathbf{v})\frac{\partial \delta(\boldsymbol{\phi}-\boldsymbol{\psi})}{\partial \boldsymbol{\phi}}\frac{\partial \boldsymbol{\phi}}{\partial t}, \quad (9)$$

then substituting $\frac{\partial \mathbf{u}}{\partial t}$ and $\frac{\partial \boldsymbol{\phi}}{\partial t}$ from Eq. 2 and Eq. 3. After some algebraic manipulations, the evolution equation for the joint PDF $f(\mathbf{u},\boldsymbol{\psi};\mathbf{x},t)$ can be written as

$$\rho\frac{\partial f}{\partial t} + \rho v_j\frac{\partial f}{\partial x_j} + (\rho g_j - \frac{\partial \langle p \rangle}{\partial x_j})\frac{\partial f}{\partial v_j} + \frac{\partial}{\partial \psi_k}[\rho S_k(\boldsymbol{\psi})f]$$
$$= \frac{\partial}{\partial v_j}[\langle -\frac{\partial \tau_{ij}}{\partial x_j} + \frac{\partial p'}{\partial x_j}|\mathbf{v},\boldsymbol{\psi}\rangle f] - \frac{\partial}{\partial \psi_k}[\langle \frac{\partial J_i^k}{\partial x_i}|\mathbf{v},\boldsymbol{\psi}\rangle f] \quad (10)$$

where the scalar flux J_i^k and the viscous force τ_{ij} have the following form:

$$J_i^k = D_{\phi_k}\frac{\partial \phi_k}{\partial x_i}, \quad (11)$$

$$\tau_{ij} = \mu(\frac{\partial u_i}{\partial x_j} + \frac{\partial u_j}{\partial x_i}). \quad (12)$$

In this equation, the mean pressure $\langle p \rangle$ can be obtained from the Poisson equation:

$$\nabla^2 \langle p \rangle = -\rho\frac{\partial^2 \langle u_i u_j \rangle}{\partial x_i \partial x_j}. \quad (13)$$

A remarkable feature of the PDF approach can be seen by observing that all the terms on the left-hand side of Eq. 10 are closed, which include the turbulent advection, the production by the mean pressure gradient, gravity and chemical reactions. However, the three conditional expectation terms on the right-hand side, which represent the transport in velocity space by viscous forces and fluctuating pressure gradient and the transport in composition space by scalar mixing, are undetermined. Modeling these terms remains challenging and is discussed in Tsai and Fox (1994).

Composition PDF Equation

The balance equation for composition PDF can be obtained by integrating the velocity-composition PDF equation (Eq. 10) over velocity space. The final form can be written as

$$\rho\frac{\partial f_\phi}{\partial t} + \rho\langle v_j \rangle\frac{\partial f_\phi}{\partial x_j} + \rho\frac{\partial}{\partial x_j}\langle v_j'\mathcal{F}\rangle + \frac{\partial}{\partial \psi_k}[\rho S_k(\boldsymbol{\psi})f_\phi] =$$
$$-\frac{\partial}{\partial \psi_k}[\langle \frac{\partial J_i^k}{\partial x_i}|\mathbf{v},\boldsymbol{\psi}\rangle f_\phi]. \quad (14)$$

Except the disadvantages when dealing with variable density (Pope, 1985), this equation is easier to model and contains no velocity variables, which reduces the computational intensity considerably. The extra unclosed terms $\langle v_j'\mathcal{F}\rangle$ can be modeled using the gradient-diffusion assumption:

$$\langle v_j'\mathcal{F}\rangle = -\Gamma_T\frac{\partial f}{\partial x_j}. \quad (15)$$

Here local isotropy is implied by Γ_T. In a more general model Γ_T should be treated as a tensor instead of a scalar (Rogers et al., 1989).

Using fractional steps for advection, diffusion and reaction in an Lagrangian frame, diffusion and reaction are treated the same way as in Tsai and Fox (1994), and advection is simulated in $\mathbf{x}$ space by the following Langevin equation:

$$\Delta_{\delta t}\mathbf{x} = \mathbf{A}\delta t + B^{1/2}\Delta_{\delta t}(W_i)_t. \quad (16)$$

The corresponding evolution equation for f_ϕ is

$$\frac{\partial f_\phi}{\partial t} + \frac{\partial}{\partial x_i}[A_i f_\phi] - \frac{1}{2}\frac{\partial^2}{\partial x_i \partial x_i}[Bf_\phi] = 0. \quad (17)$$

In order for this equation to evolve according to the gradient-diffusion assumption, the coefficients should be written as

$$B = 2\Gamma_T/\rho \quad (18)$$

and

$$\mathbf{A} = \langle \mathbf{u} \rangle + \nabla\Gamma_T/\rho. \quad (19)$$

Therefore, during the simulation, the new position of each Lagrangian particle is obtained from

$$\begin{aligned}\mathbf{x}(t+\Delta t) &= \mathbf{x}(t) + [\langle \mathbf{u} \rangle + \nabla\Gamma_T/\rho]_{\mathbf{x}_t}\Delta t \\ &+ [2\Delta t\Gamma_T/\rho]_{\mathbf{x}(t)}^{1/2}\boldsymbol{\xi}\end{aligned} \quad (20)$$

where $\boldsymbol{\xi}$ is a standardized joint normal random vector. The value of Γ_T is derived by comparing the gradient-diffusion assumption with the $k-\epsilon$ model (Launder and Spalding, 1974). The turbulent viscosity μ_t in the $k-\epsilon$ model is defined as $C_\nu k^2/\epsilon$. A similar definition can be used for scalar turbulent diffusivity:

$$\Gamma_T = \mu_t/\sigma_\phi \quad (21)$$

where σ_ϕ is the turbulent Schmidt number taken to be 0.7. As with the velocity-composition PDF, the overall yield is insensitive to the value of Γ_T.

NUMERICAL SIMULATION

In Li and Toor's experiment, the single-jet reactor consisted of a smaller tube centered with four vanes in a larger tube. The smaller tube had a 0.004 m inner diameter and 0.0048 m outer diameter. The larger tube had an inner diameter of 0.0066 m and a length of 1.83 m. As reported by Li and Toor (1986), the large-scale radial concentration gradients were gone in about 0.04 m at $Re = 2,300$ and 0.006 m at $Re = 7,000$, after which the reaction was essentially one dimensional. The computational domain in the simulation was set to be long enough for reactions to go to completion. Flow fields with two different Reynolds numbers (3,530 and 7,552) were simulated. A length of 0.5 m for $Re = 7,552$ and 0.4 m for $Re = 3,530$ were used. Although the large-scale radial gradient disappears faster at higher Reynolds number, the higher mean velocity forces the chemical reaction to go further down the tube.

The inlet conditions were assumed to be fully developed annular pipe flow. In both cases ($Re = 3,530$ and $7,552$), the flow fields in the upstream outer annular pipe were laminar ($Re < 2,100$ based on the hydrodynamic diameter and mean velocity) and a moderate turbulent intensity of 0.04 was set as the inlet condition. There were 68×23 cells in the simulation domain at $Re = 3,530$ and 68×16 cells at $Re = 7,552$. In the figures the cell numbers are doubled using linear interpolation for smoother output. Both Reynolds numbers have nonuniform grid spacing in order to fit the grids with the inner tube wall at the inlet and to give appropriate cell size at the near wall regime. The grid spacing in the axial-direction is stretched by a factor of 1.04 and 1.05 for the flow fields at $Re = 3,530$ and $Re = 7,552$, respectively.

The PDF codes are initialized by reading the steady-state solution of the flow field at each grid from the output of FLUENT, including mean velocity (u_x and u_y), k, ϵ and mean static pressure $\langle p \rangle$. Then the averaged values of those quantities are linearly interpolated onto each stochastic particle according to its location. At each time step, new notional particles are initialized from the inlet with proper weights, flow field properties and scalar concentrations. The number of particles in each cell is controlled by separating heavy particles into lighter particles or clustering lighter particles into heavier particles according to the average notional particle weight in each cell.

The velocity and scalar components are updated by solving two stochastic differential equations (Tsai and Fox, 1994). The solutions of these two equations can be written, respectively, as

$$\begin{aligned} v_{i,t+\Delta t}^* &= (v_{i,t}^{*\prime} - \langle v_i^* \rangle) e^{-C_L(\epsilon/k)\Delta t} \\ &+ \sqrt{C_0 \epsilon^2 (1 - e^{-2C_L \Delta t})/(2kC_L)} \xi_i \\ &- \frac{\partial \langle p \rangle}{\partial x_i} \Delta t + \langle v_i^* \rangle \end{aligned} \quad (22)$$

and

$$\psi_{i,t+\Delta t}^* = (\phi_{i,t}^* - \langle \psi_i^* \rangle) e^{-0.5 \frac{C_\phi}{\tau_\phi} \Delta t} + \langle \psi_i^* \rangle, \quad (23)$$

where Δt is the time step, $\langle v_i^* \rangle$ and $\langle \psi_i^* \rangle$ are the local mean values of v_i and ψ_i, derived by taking the ensemble average of the particle values of v_i^* and ψ_i^* in each cell, ξ_i is a Gaussian random number with

$$\langle \xi_i \xi_j \rangle = \delta_{ij} \quad (24)$$

and

$$C_L = \frac{1}{2} + \frac{3}{4} C_0, \quad (25)$$

where $C_0 = 2.1$ (Pope, 1985).

The changes of scalar compositions due to the chemical reaction are calculated through the linear interpolation of a look-up table, which contains the values of composition changes in one time step for scalars A, B, R and S with three independent parameters (A, B and R) using a uniform grid ($50 \times 50 \times 50$). The values of this table were computed by a stiff equation solver (LSODAR) developed by Petzold and Hindmarsh (1983). The accuracy of using the look-up table has been tested by using different numbers of grids and compared with direct integration using a 4th-order Runge-Kutta method.

RESULTS AND DISCUSSION

Li and Toor (1986) performed extensive experiments on single- and multi-jet tubular reactors with different Reynolds numbers and scalar concentrations. Our simulation results are compared with the experiments for the single-jet reactor. The accuracy of different mixing rate models is verified through comparisons with the yield of product R. The validity of a mixing rate model depends on its ability to predict the correct yield of R at different Reynolds numbers once the mixing rate constant C_ϕ is determined. A good mixing rate model should have a C_ϕ independent of Reynolds number and scalar concentration. Although the Schmidt number is an important factor that may affect the value of C_ϕ, it is a property of chemical species and assumed to be the same for different species in this reaction as indicated by Li and Toor (1986). Therefore it can be treated as a single factor included in C_ϕ. The determination of C_ϕ was done by fitting the simulation results with experimental data for one combination of Reynolds number and initial scalar concentrations (Tsai and Fox, 1994). The same value of C_ϕ was used for subsequent simulations where the Reynolds number and initial scalar concentrations were varied independently.

The detailed comparison between the simulation and the experimental data has been reported elsewhere

(Tsai and Fox, 1994). Here, only the results from the two different PDF codes (LVCPDF and LCPDF) are compared. Tables 1 and 2 present the simulation results and experimental data in terms of the yield defined by

$$Y_R = \frac{\langle \phi_R \rangle}{\langle \phi_R \rangle + 2\langle \phi_S \rangle}. \quad (26)$$

The simulation values were obtained by taking the average of the particle values at the outlet over 20 time steps. From the results, it can be seen that both methods predict nearly identical yields. In most cases, the difference is less than 1.0 %. This confirms that the gradient-diffusion assumption works well for this type of reactor. The excellent agreement makes the LCPDF code a good candidate for other incompressible flow applications.

The distributions of scalar fields A and B at $Re = 3{,}530$ with species B injected through the center tube at a concentration of 2.86×10^{-3} kmol/m^3 found using the LCPDF code are shown in Figs. 1–2. These results are very close to the predictions of the LVCPDF code shown in Figs. 3–4. The one notable difference is the rate of scalar flux in the radial direction (cf. Figs. 3 and 1). With the LVCPDF code ($C_0 = 2.1$) the radial flux is higher so that, for example, reactant A reaches the centerline faster than with the LCPDF code. Nevertheless, as pointed out earlier, this difference has no effect on the overall yield (Tables 1 and 2).

CONCLUSIONS

Using the fact that no model is required for turbulent advection in the LVCPDF code, a LCPDF code with the gradient-diffusion assumption has been validated for incompressible flows. As shown in this study, both methods give almost the same results for Li and Toor's experiment. Because the LCPDF code contains less variables, it is the more computationally tractable choice for simulating incompressible turbulent reacting flows.

Although better scalar mixing models are still necessary to improve their predictions, both the LVCPDF and the LCPDF codes are useful and practical methods for studying the effects of turbulence on complex chemical reactions. Further extension of these two PDF codes to 3-dimensional problems with liquid-phase reactions is straightforward. The average computational time for the simulations reported in this work is about 24 hours for the LVCPDF code and 6 hours for the LCPDF code on a HP-Apollo 735 workstation. Thus, a 3-dimensional simulation with 10 to 20 times more grids should be still within the reach of modern supercomputers. Another advantage of employing PDF methods is that the Monte-Carlo technique can be easily implemented by parallel algorithms (Pope, 1992). By exploiting these techniques on a multiprocessor computer, PDF methods should prove even more affordable and will be of great interest for both academic research and industrial applications.

NOTATION

$\mathbf{A}$	linear drift tensor in the Langevin equation for the joint LCPDF equation
B	coefficient of the Wiener process in the Langevin equation
C_0	universal constant in simple Langevin model
C_ϕ	empirical constant for scalar micromixing model
C_L	$= \frac{1}{2} + \frac{3}{4}C_0$
D_ϕ	diffusivity constant of ϕ
d	inner diameter of the reactor
f	probability density function
g_i	gravity
J	scalar flux
k	turbulent kinetic energy
k_1	reaction rate constant for first reaction step
k_2	reaction rate constant for second reaction step
p	pressure
S_i	chemical source term of species i
t	time
u_i	velocity component
v	phase variable of u
v^*	random variable of v
W_i	isotropic Wiener process
x_i	spatial variable
$\mathbf{x}$	position vector
Y_R	yield of product R

Greek Variables

Δt	time step
$\mathcal{F}$	fine-grained density function
δ_{ij}	Dirac delta function
ϵ	turbulent energy dissipation
Γ_T	eddy diffusivity
ϕ	concentration
ψ	phase variable of ϕ
ψ^*	random variable of ψ
ρ	density
$\langle \psi \rangle$	expected value of ψ
σ_ϕ	turbulent Schmidt number

τ	integral time scale
τ_ϕ	scalar mixing time scale
τ_{ij}	viscous force
μ	viscosity
μ_t	turbulent viscosity
ν	kinematic viscosity
ξ	standardized joint normal random vector

ACKNOWLEDGEMENTS

The authors would like to thank Prof. S. B. Pope for allowing us to access his LVCPDF code *PDF2DS*. This work was supported by the National Science Foundation under grant CTS-9158124 (PYI Award) and The Dow Chemical Company.

LITERATURE CITED

1. Anand, M. S. and S. B. Pope, *Turbulent Shear Flows 4* (L. J. S. Bradbury et al., Eds., Springer-Verlag), 46 (1984).

2. Dopazo, C., "Probability density function approach for a turbulent axisymmetric heated jet. Centerline evolution," *Phys. Fluids* **18**, 397 (1973).

3. Gardiner, C. W., 1990, *Handbook of Stochastic Methods* (2nd Edn.), Springer-Verlag, New York.

4. Li, K. T. and H. L. Toor, "Turbulent reactive mixing with a series-parallel reaction: Effect of mixing on yield," *AIChE J.* **32**, 1312 (1986).

5. Lundgren, T. S., "Distribution functions in the statistical theory of turbulence," *Phys. Fluids*, **10**, 969 (1967).

6. Launder, B. E. and D. B. Spalding, "The numerical computation of turbulent flows," *Comput. Methods Appl. Mech. Eng.* **3**, 269 (1974).

7. Monin, A. S. and A. M. Yaglom, *Statistical fluid mechanics*, (MIT Press, Cambridge, Mass, 1971).

8. O'Brien, E. E., *Turbulent Reactive Flows* (P. A. Libby and F. A. Williams, Eds, Springer-Verlag, Berlin, 1980), Chap. 5.

9. Petzold, L. R. and A. C. Hindmarsh, *LSODAR: Livermore solver for ordinary differential equations, with automatic method switching for stiff and non-stiff problems, and with root finding* (1987).

10. Pope, S. B., "Pdf methods for turbulent reactive flows," *Prog. Energy Combust. Sci.* **11**, 119 (1985).

11. Pope, S. B., *PDF2DS: A code to solve the velocity-composition joint pdf equation,* unpublished FORTRAN code listing (1991).

12. Pope, S. B., "PDF/Monte Carlo methods for turbulent combustion and their implementation on parallel computers," presented at the Sixth Toyota Conference, Japan, October (1992).

13. Rogers, M. M., N. N. Mansour and W. C. Reynolds, "An algebraic model for the turbulent flux of a passive scalar," *J. Fluid Mech.* **203**, 77 (1989).

14. Tsai, K. and R. O. Fox, "PDF simulation of a turbulent series-parallel reaction in an axisymmetric reactor," *Chem. Eng. Sci.* (to appear).

15. Tsai, K. and E. E. O'Brien, "A hybrid one- and two-point approach for isothermal reacting flows in homogeneous turbulence," *Phys. Fluids A* **5**, 2901 (1993).

Table 1: Yield of the series-parallel reaction produced by the experiment of Li and Toor (1986) and the LVCPDF code using $C_\phi \sqrt{k}/d$ as the scalar mixing rate. († C_ϕ value determined using this case.)

		Experiment		*Simulation* ($\sqrt{k}/d$)		
Conc. of B gmol/m^3	Reynolds No.	Jet is Reactant B	Jet is Reactant A	Jet is Reactant B	Jet is Reactant A	C_ϕ
2.86	3,530	82.05%	84.92%	81.0%	80.5%	1.65
2.86	7,552	88.33%	90.68%	88.3% †	88.38%	1.65
0.964	3,530	88.95%	91.13%	91.8%	90.0%	1.65
0.964	7,552	93.45%	94.88%	95.4%	95.3%	1.65

Table 2: Yield of the series-parallel reaction predicted by the LCPDF code using $C_\phi \sqrt{k}/d$ as the scalar mixing rate with $C_\phi = 1.65$.

Simulation ($\sqrt{k}/d$)				
Conc. of B gmol/m^3	Reynolds No.	Jet is Reactant B	Jet is Reactant A	C_ϕ
2.86	3,530	80.98%	79.85%	1.65
2.86	7,552	88.5%	88.11%	1.65
0.964	3,530	91.78%	90.72%	1.65
0.964	7,552	95.35%	95.43%	1.65

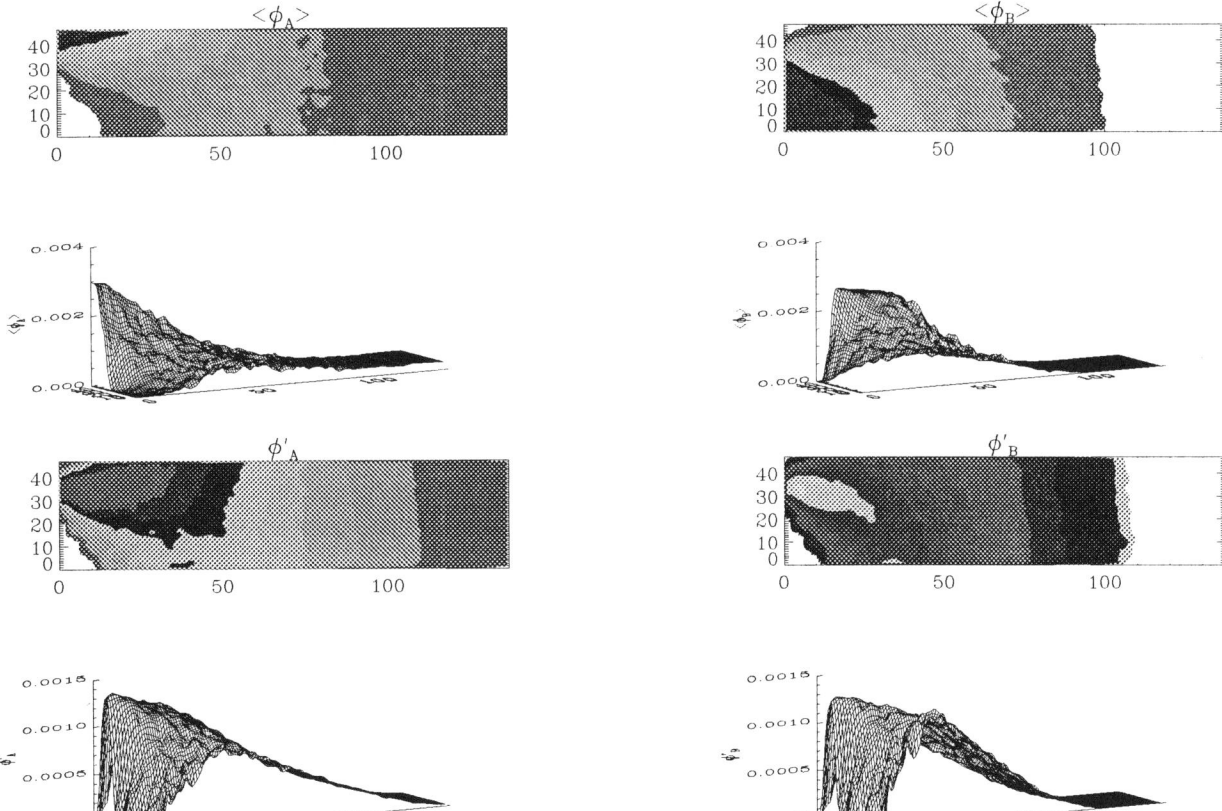

Figure 1: Distribution of the mean (upper) and standard deviation (lower) of A predicted by the LCPDF code at $Re = 3,530$.

Figure 2: Distribution of the mean (upper) and standard deviation (lower) of B predicted by the LCPDF code at $Re = 3,530$.

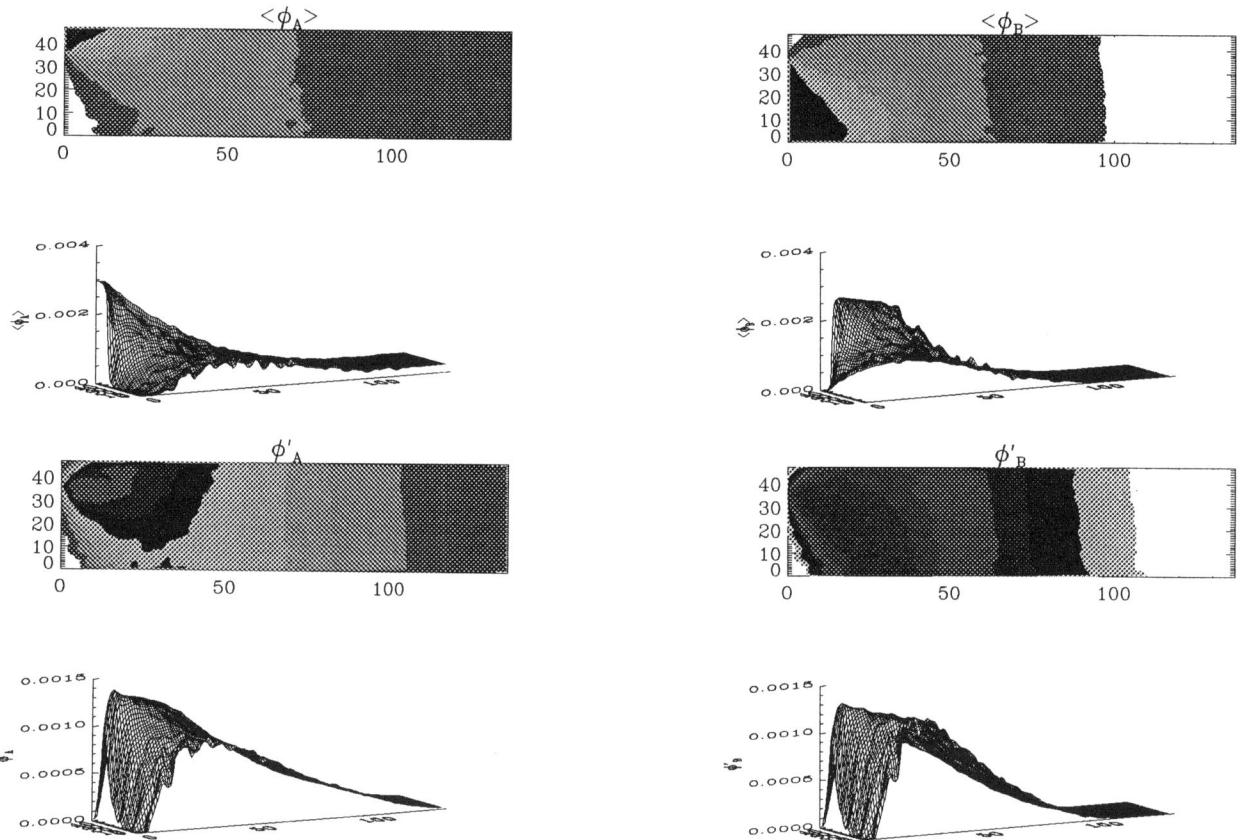

Figure 3: Distribution of the mean (upper) and standard deviation (lower) of A predicted by the LVCPDF code at $Re = 3,530$.

Figure 4: Distribution of the mean (upper) and standard deviation (lower) of B predicted by the LVCPDF code at $Re = 3,530$.

A Comparison of Experimental Data and Numerical Simulation of Mixing in Forced-Circulation Evaporative Crystallizers

Paul A. Gillis and Tim W. Gambrel
The Dow Chemical Company, Texas Operations, B-1226, Engineering Sciences,
2301 Brazosport Boulevard, Freeport, TX 77541

Proper mixing and efficient phase-disengagement strongly influence the performance of forced-circulation evaporative crystallizers. Crystallizer models can provide useful insight into determining the effect of operational and design modifications on sustainable capacity, fouling rates, and the product's crystal size distribution. Laser sheet illumination and laser doppler velocimetry were used to characterize the flow in a scaled-down evaporator geometry. Mixing in this industrial crystallizer geometry was simulated with computational fluid dynamics. Comparisons between the experimental data and numerical predictions showed general agreement when sufficient grid resolution is employed.

Computational fluid dynamics and experimental studies are combined to evaluate the effect of design and operational modifications on the performance of industrial crystallizers. This study focuses on understanding the role of mixing in forced-circulation evaporative crystallizers. Mixing significantly impacts mass transfer, heat transfer, phase separation, and the crystal size distribution (CSD) of the product.

Forced-circulation crystallizers employ a strong recirculating flow to provide agitation for the crystallizer. Due to the height of these vessels, hydrostatic pressure restricts boiling to a region near the liquid free surface. The principal production objectives are to maximize process capacity, minimize fouling in the evaporator, and produce large and narrowly-distributed crystals. A comprehensive model was developed by integrating relevant phenomenological sub-models into a computational fluid dynamics (CFD) model. This model is useful in simulating the interaction of key crystallization phenomena.

Previously crystallization modeling work has focused primarily on global process models for mixed-suspension mixed-product removal (MSMPR) crystallizers [1,2]. These modeling tools predict the CSD resulting from specified growth kinetics, residence times, and initial seeding conditions. These models assume that the contents of the crystallizer are fully mixed.

Additional prior work has addressed the effect of vessel geometry on the crystallization process [3]. The importance of mixing and agitation in crystallizers has also been studied [4]. This work differs through the application of computational fluid dynamics to predict the effect of vessel configuration on mixing and through the combination of additional process models with the CFD model.

Velocity data was measured in a scaled-down evaporator geometry. Several flow simulations were converged and compared to the vertical velocity data collected by a laser doppler velocimetry (LDV). The simulated and measured flow fields were essentially in agreement.

CRYSTALLIZATION PROCESS

Forced-circulation evaporative crystal-lizers, also known as evaporators, are used in the manufacture of numerous chemicals. Evaporative crystallization processes can become fairly large reaching production rates of 100,000's lb/hr and equipped with recirculation loop piping as large as 6 feet in diameter.

A typical industrial evaporative crystallizer geometry is depicted in Figure 1. The main flow is driven by an axial pump that recirculates flow through a heat exchanger and evaporator body. Liquid inlet and slurry outlet streams can be placed along the recirculation loop. Most of the crystallization and phase separation occur within the evaporator body. This paper will focus on the mixing in the evaporator body.

Capacity, reliability, and product quality are key issues that determine the performance of an evaporative crystallizer. Capacity is influenced by the maximum sustainable production rate as well as the frequency and duration of water treatments to remove encrustation build-up. A typical measure of product quality is the crystal size distribution of the solids.

Two factors that affect production rate are vapor/liquid disengagement and heat transfer rates. Phase separation is mainly affected by liquid inlet location and design, recirculation flow rate, and boiling rate. The heat transfer rates are influenced by the temperature drop in the evaporator and the recirculation rate.

The size distribution of the crystal product is influenced by a large number of factors. Important factors include the mixing in the boiling zones as well as the concentration of existing crystals in these "production zones." Crystal size can also be influenced by mechanical contacting and fluid shear rates.

CRYSTALLIZER MODELING

A comprehensive crystallization model was developed to better understand the various phenomena occurring within an evaporator as well as how these phenomena interact [5]. Due to the importance of mixing in these vessels, the model was based on a computational fluid dynamics code. Additional sub-models have been developed to address phase separation, crystal suspension, production, and crystal growth.

An example of the model's capabilities is shown in Figure 2. This figure illustrates the coupling between pressure, temperature, and production rate. The pressure field (Figure 2A) predicted by the CFD model includes hydrostatic and dynamic pressure contributions. In this example, a high temperature fluid is introduced in the inlet and flows up the riser pipe towards the free surface. When the fluid enters a computational location where the pressure is lower than its vapor pressure, boiling is predicted. The simulated temperature field (Figure 2B) is a result of maintaining the fluid at or below its bubble point. Local production rates are determined from the temperature field. This is accomplished by relating the temperature gradients to boiling rates and assuming (for a high-yield Class II system) that supersaturation is immediately relieved by crystal formation and growth.

The interaction of the various sub-models in the overall crystallizer model has revealed competing mixing objectives. In order to satisfy some of the design objectives, it is desirable to have good mixing (recirculating, shearing flows), while for other objectives it is best to have poor mixing (smooth, "plug" flows).

For example, poor mixing is advantageous for phase separation as it allows the buoyant forces to overcome drag. However, in this region of the crystallizer (the boiling zone), good mixing is needed to de-localize supersaturation and to ensure the suspension of existing crystals.

EXPERIMENTAL SETUP

It is important to evaluate the accuracy of theoretical models with experimental data. For this purpose, a laboratory flow rig was constructed. Ideally, model components should be evaluated individually. The laboratory apparatus was designed to allow for comparisons of both single and multi-phase flows. The design of the apparatus was based on a scale-down of a possible configuration for an industrial evaporator.

An overview of the flow rig with dimensions is given in Figure 3. The main cylindrical body is constructed of transparent acrylic to provide visual observation and laser access to internal flows. The flow rig is connected to a fluid storage tank and centrifugal pump with 3 inch pipe. Water (90 gpm) is introduced in the vessel and flows upward through the riser pipe until it strikes a conical flow deflector (top-hat). The gas/liquid interface is maintained below the top outlet.

NUMERICAL RESULTS

Flow within the lab rig was modeled using Fluent™ 4.25 CFD software from Fluent, Inc. A single vertical plane of symmetry, corresponding to the plane shown in Figure 3, was used to reduce the number of grid points. The final single-phase predictions were generated on a 149,688 node grid. A lower portion of the body-fitted-coordinates grid is shown in Figure 4. The gray sections within this grid denote internal walls. A difficulty with using the hexagonal elements characteristic of a finite-volume (finite-difference) method is the handling of conical (or triangular) shapes. This resulted in considerable grid skew near the bottom of the flow deflector.

Figure 5 contains velocity predictions for the symmetry plane. The upward velocity in the riser pipe is approximately 4 ft/s. The vectors symbolize the predicted magnitude and direction of the fluid at the location specified by the end of the vector tail.

The model predicts a strong jet deflected horizontally by the top-hat. This jet strikes the vessel wall and creates recirculating flows above and below the flow deflector. The flow through the liquid outlet can be seen in the lower left part of the plot. All the flow features predicted in this plot have been experimentally verified with laser sheet illumination.

Figure 6 shows a close-up view of predicted velocity vectors around the top-hat. This plot focuses on the two recirculation zones created by jet impingement on the wall. These predictions are in a vertical plane perpendicular to Figure 5 and pass through the central axis of the cylindrical vessel. Two gray rectangles were placed in this plot to denote the location where experimental velocity data were collected.

COMPARISON OF RESULTS

Velocity measurements were obtained using laser doppler velocimetry (LDV). A single component system with a 300 mW Argon Ion laser was used to measure the vertical component of velocity. Approximately two thousand instantaneous velocities were measured at each location. Velocities were mapped in the vertical plane seen in Figure 6. Comparisons are presented for the two lines of data at the locations depicted by the gray boxes in Figure 6.

The first comparison is for the data located 65 mm above the top edge of the flow deflector. Figure 7 contains experimental data and numerical predictions for this location. The solid circles denote the average upward velocity measured with LDV. Two separate LDV data sets are included in these comparisons. The bars above and below the LDV data represent one standard deviation from the mean based upon the instantaneous velocity measurements.

Flow predictions for two different grid resolutions (denoted COARSE and FINE) are provided in Figure 7. The coarse grid (14,448 nodes) simulation under-predicted the strength of the recirculation zone above the top-hat. Increasing the grid density to 149,688 nodes improved the agreement, but the velocity magnitudes remained under-predicted. It is possible that further refinement of the grid could result in closer agreement between the experimental and modeling results.

Figure 8 contains the comparison of velocities below the top-hat. In this figure, only the finer computational grid is used. The velocity magnitudes are matched well near the wall and near the riser pipe. The overall agreement is better than in Figure 7. The biggest discrepancy is the prediction of the center of the recirculation zone. The LDV data indicates that the zero velocity location is approximately 20 mm closer to the wall than predicted by the CFD simulation.

Two turbulent models are also compared in Figure 8. These models were the standard k-ε model (KEPS) and the re-normalized group (RNG) k-ε model. The RNG model produced a slight improvement in agreement with the experimental data.

CONCLUSIONS

Industrial evaporative crystallization is a complex process. Modeling provides a fundamental understanding of the various important phenomena and how these phenomena interrelate. A laboratory flow rig was used to generate data for comparing with single-phase flow simulations. Laser sheet illumination confirmed the overall flow patterns predicted with a CFD model. Velocity predictions were compared with LDV data and general agreement was observed when sufficient grid resolution was employed.

ACKNOWLEDGMENTS

The authors are grateful for the efforts of Ravi Shanker in providing expertise in the use of the laser doppler velocimeter for the collection of velocity data presented in this paper. We would also like to acknowledge the efforts of Bob Spradling for his efforts in preparing and running the laboratory apparatus.

LITERATURE CITED

Randolph, A.D. and M.A. Larson, *Theory of Particulate Processes*, Academic, New York/London, 2nd edition (1988).

Mydlarz, J. and A.G. Jones, "On the estimation of size-dependent crystal growth rate functions in MSMPR crystallizer," The Chemical Engineering Journal, **53**, 125 (1993).

Bennett, R.C., "Crystallizer Selection and Design," *Handbook of Industrial Crystallization*, Allan S. Myerson, ed., Butterworth-Heinemann (1992).

Oldshue, J.Y., "Agitation and Mixing," *Handbook of Industrial Crystallization*, Allan S. Myerson, ed., Butterworth-Heinemann (1992).

Gillis, P.A. and T.W. Gambrel, "Numerical Simulation of the Effect of Crystallizer Hydrodynamics on Crystal Concentrations, Salt Production and Vapor-Liquid Disengagement," presented at the 4th Annual Meeting, Association of Crystallization Technology (1994).

42 Industrial Mixing Fundamentals with Applications

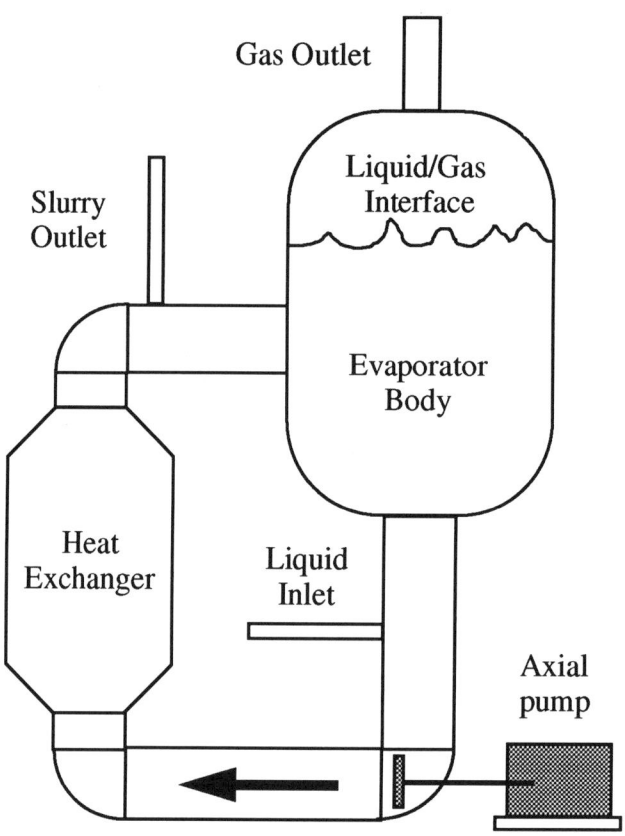

Figure 1. Industrial Evaporator Geometry

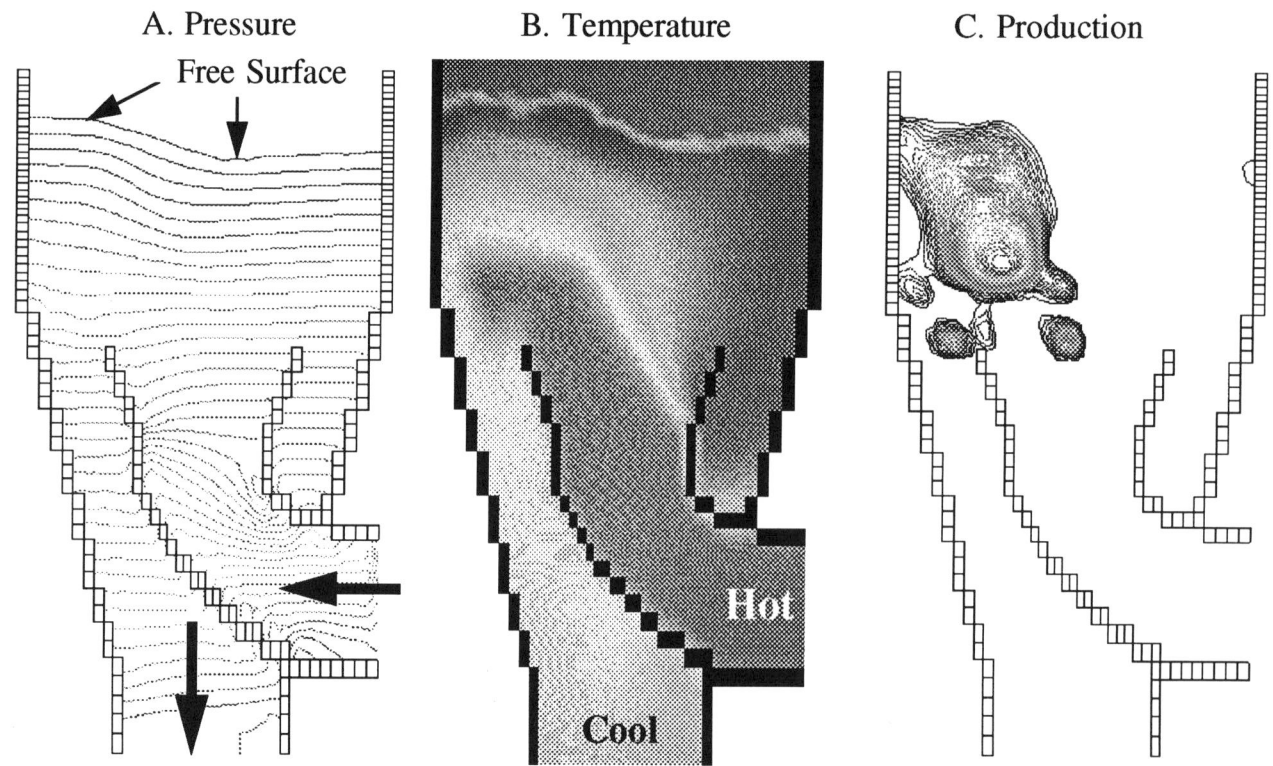

Figure 2. Coupling Between Pressure, Temperature, and Production

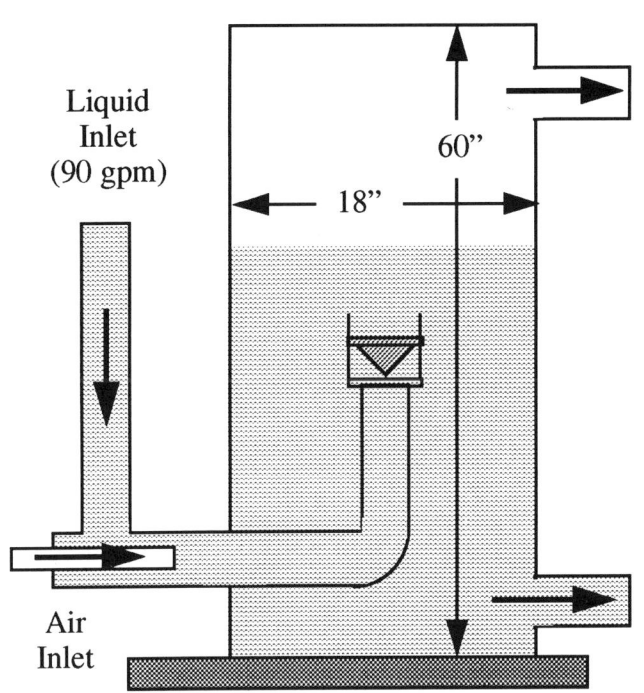

Figure 3. Laboratory Flow Apparatus

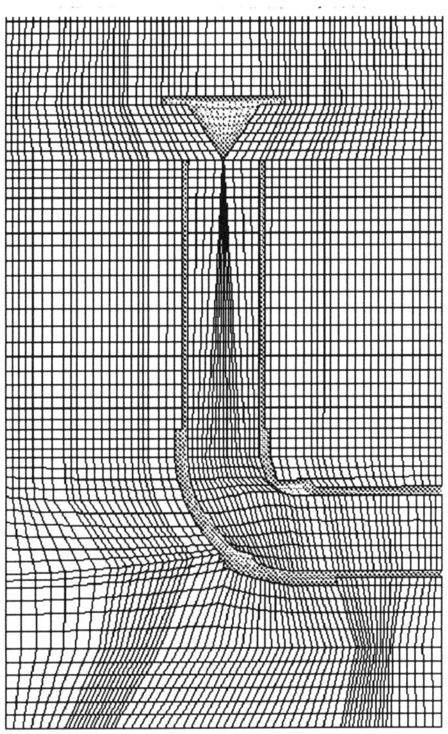

Figure 4. Computational Grid

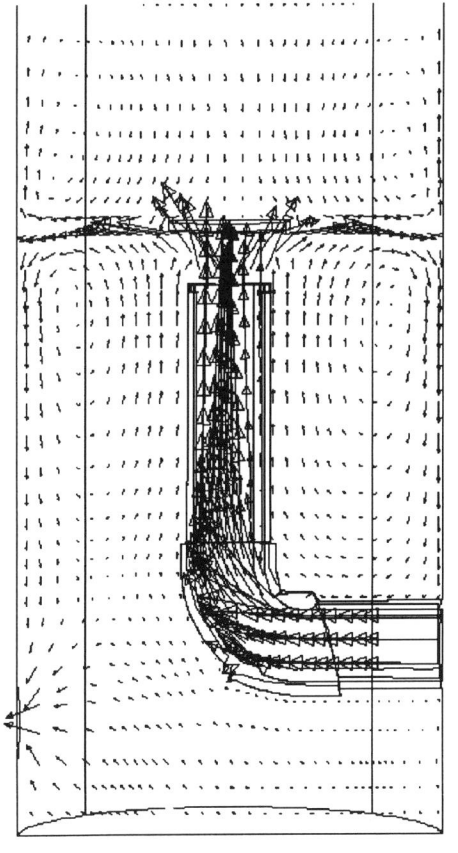

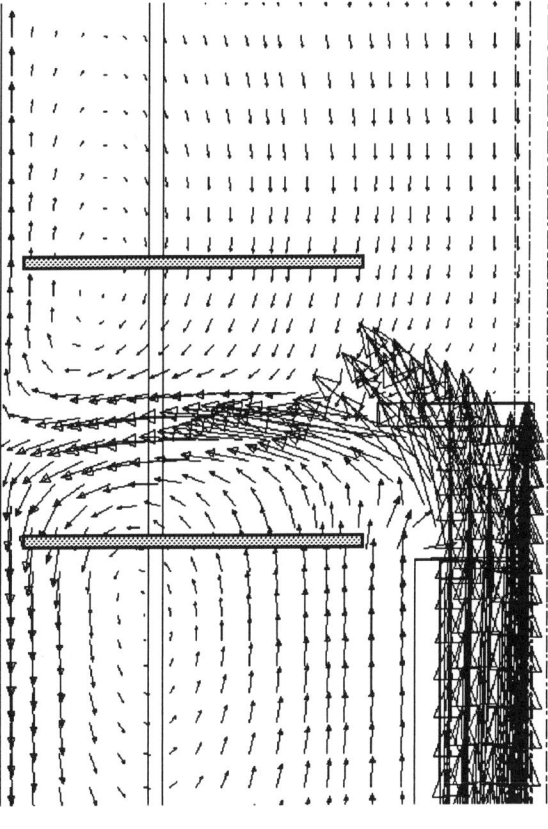

Figure 5. Flow Field Simulation

Figure 6. Predicted Velocities Near Top-Hat

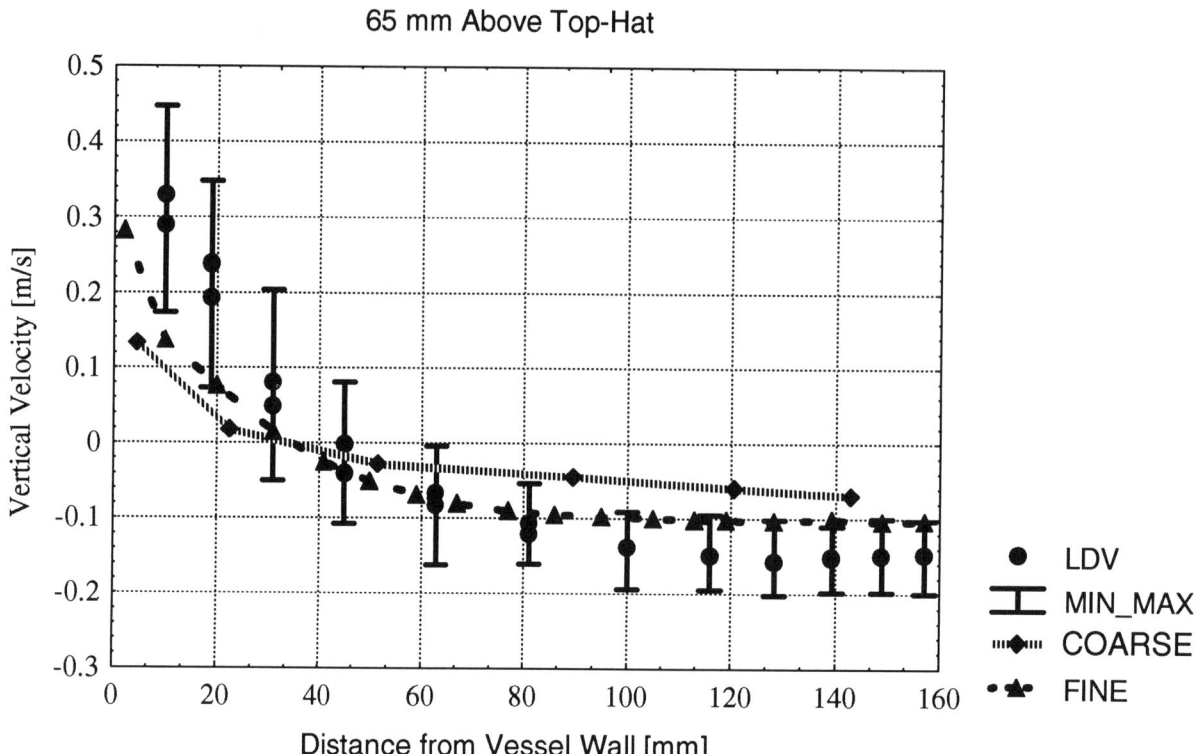

Figure 7. Velocity Comparison Above Top-Hat

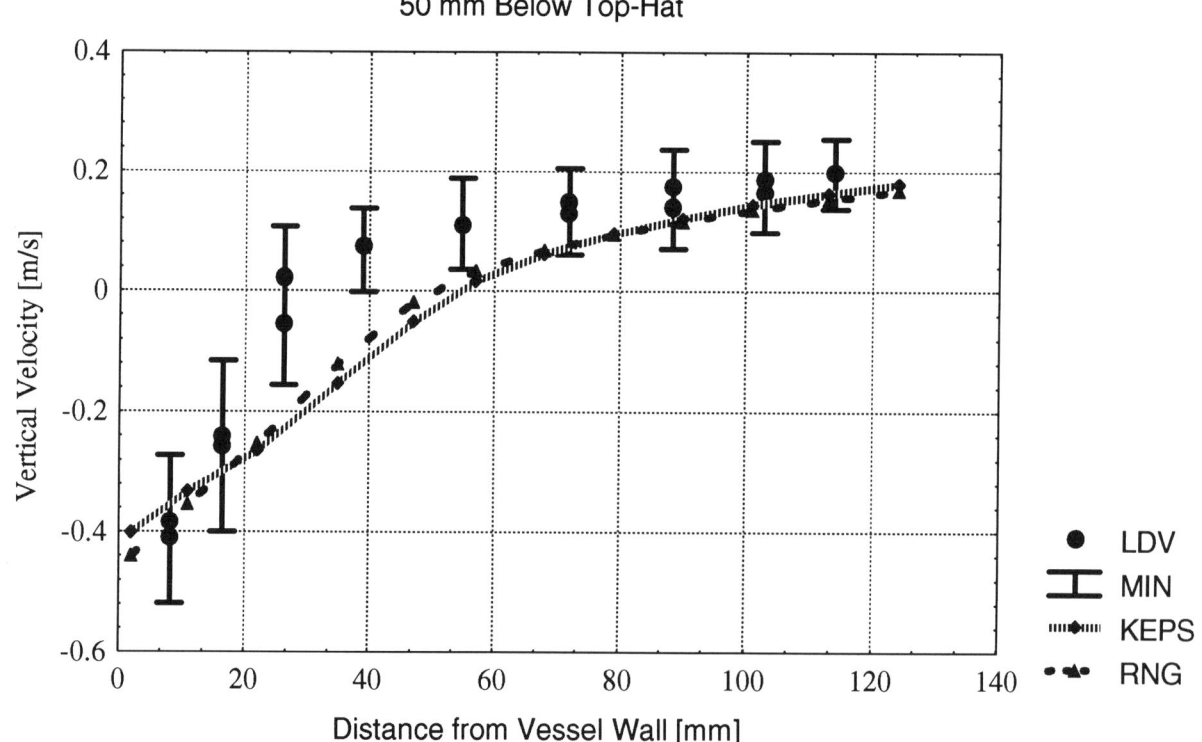

Figure 8. Velocity Comparison Below Top-Hat

A Relationship Between Grinding in Fluidized-Bed Units as a Function of Reynolds Number

Robert D. Knecht, Matthew Hazleton
Chemical Engineering and Petroleum-Refining Engineering Department
Colorado School of Mines, Golden, CO 80401

Stevan P. Stiefvater
EG&G Rocky Flats, Inc., P.O. Box 464, Golden, CO 80402

Bubbling mechanisms in fluidized-bed systems maintain a homogenous mixture of particles throughout the bed. Bubbles explode at the surface of the bed and spray solids into the freeboard. Finer particles carry from the bed, based on the particle size distribution, flow conditions and bed geometry. Carry over is constrained by the saturation capacity of the gas stream, which is a function of the solid particles and flow conditions. Solids carry over decreases bed hold up and reduces the stability of the bed. An attrition study was conducted in a fluid-bed, flow visualization minipilot plant to analyze particle distribution under various operating conditions and to correlate the distribution as a function of material properties and fluidizing conditions. This study focuses on an unsteady-state evaluation of bed conditions, in which feed is not added to the bed and solids only leave the bed through entrainment.

Grinding, which occurs as particles rub against each other, produces shear forces that cause particles to abrade, and these particles distribute over a range of sizes regardless of the size distribution of the feed material. Mechanical energy created by the flow characteristics causes grinding of the bed media, and regression analysis describes the fraction of fine materials as a power function of particle size which depends exponentially on time.

Fluid-bed technology has developed tremendously since it began around World War II. Mullen (1) noted that this technology was applied to combustion processes beginning in the early 1960's, and since then over 300 units, installed worldwide, have incinerated a number of chemical and petroleum products.

The Rocky Flats Plant (RFP), one of several facilities in the United Stated nuclear weapons complex, generated over 62 combustible mixed-waste streams, and research in the late 1970's and early 1980's at RFP assessed the potential of a fluid-bed thermal-treatment process to reduce the volume of transuranic waste stored at or shipped from the facility. The primary goal of the process was not only to reduce volume by removing chemically hazardous components but also to maintain control of radioactivity. The fluid-bed unit consisted of a two-stage, low-temperature treatment process followed by off-gas filtration. Thermal treatment occurred in a series of bubbling-beds, fluidized by air and nitrogen. The first treatment stage pyrolyzed waste and neutralized chloride gases, and the gas stream passed through a cyclone to remove solids. The second stage utilized a catalytic combustion technology to complete thermal destruction of the waste, and finally, the gas stream passed through a series of filters to eliminate any remaining solids from the gas stream.

Bed media consisted of a mixture of catalyst to improve kinetics at low temperatures and sorbent (sodium carbonate) to neutralize chloride off gases. Particles of salt (NaCl), which formed as sodium carbonate neutralized the gases, created an "armor-like" barrier on the sorbent particle. Mass transfer mechanisms for neutralization relied on removal of this outer surface from the particles. Frounfelker (2) described an application in which sodium bicarbonate released water and carbon dioxide gases and produced sodium carbonate. Baking soda particles underwent a "popcorn" effect, which reduced "armor" formation, as these gases exited through the surface of the particles.

In 1990 the program resumed in response to advantages that the technology offered. A Flow Visualization Unit (FVU) was constructed at the Colorado School of Mines to study the hydrodynamic behavior of this system at ambient conditions and to serve as a tool to demonstrate fundamental hydrodynamic concepts to various interested parties. The purpose of the Attrition Study was to analyze particle attrition in the FVU following these hydrodynamic studies. Attrition occurred primarily as a result of fluidization, since the bed operated at ambient conditions (no thermal effects) without reaction (no chemical effects). The studied focused on an analysis of the fines, carry-over material, at various velocities and operating times.

BASIS

Fluidizing velocity imparts energy to the media and causes a reduction in particle size as a function of time, attrition, and changes in the physical properties of bed materials resulting from the comminution process are described in terms of the particles size distribution. According to Beke (3), a probability function best describes the passage (or retention) of a particle through a sieve (mesh). The frequency (cumulative distribution) adequately represents the distribution of particles between two size fractions. Gaudin (4) observes that many approaches have been used to describe grinding as a function of size; as an example, one approach characterizes a linear relationship for the distribution of fine particles and describes the fraction of material retained (distribution) in terms of the particle size.

$$p(r) = C_1 D_p^{C_2} \qquad 1$$

The coefficients (C_1 and C_2) characterize the maximum possible size of the bed material and relative level of grinding.

Several authors have described the attrition mechanism as an integral component of an overall model for particle size reduction in fluid-bed systems. Based on a material balance around the bed, Levenspiel, Kunii and Fitzgerald (5) and Mauri and Stecconi (6) develop separate models which simulate the particle size distribution. Both models reduce to a similar description for physical attrition in the FVU:

$$\frac{R(r)}{p_b(r)} \frac{\partial p_b(r)}{\partial r} = -\frac{3\partial r}{r \partial t} - \frac{\partial R(r)}{\partial r} - \frac{F_2 p_2(r)}{W p_b(r)} \qquad 2$$

The numerical model describes the change in particles size as a function of time, particle density and size, gas density, velocity and bed configuration. Assuming carry over from the bed is negligible, we generate a dimensionless relationship between size distribution, particle Reynolds Number, dimensionless time, and the ratio of particle to gas densities:

$$\frac{W(D_p)}{W_o} = f(N_{Re}, \theta, \frac{\rho_p}{\rho_g}) \qquad 3$$

Grinding produces a distribution which varies uniquely as a function of particle size and decreases with time.

TEST PROCEDURE

This study evaluated attrition of sodium sesquicarbonate, a mixture of sodium carbonate and bicarbonate considered an alternate sorbent for the neutralization process. Objectives of the study were to assess attrition for single component and multicomponent media and to develop a correlation which described the size distribution in the bed. We conducted an attrition test over several hours at constant velocity and operating conditions to determine the change in media properties and examined the change in the weight fraction retained by standard ASTM sieves for samples of bed media, cyclone underflow, and bag house underflow. Particle size and density data were refined to determine mean particle diameter and attrition rates. Initially we independently analyzed change in size fraction as a function of time or particle size and, then, analyzed the change as a function of both parameters using regression analyses. We produced a correlation which described the distribution of particles as a function of the particle Reynolds Number and dimensionless (particle) time.

DISCUSSION OF RESULTS

The Hydrodynamic Investigation, conducted in 1993, established baseline characteristics for the original fluid-bed configuration tested at RFP and produced data to explore attrition rates. During the period of the investigation, gas velocity averaged less than 0.2 m/s; but briefly exceeded 0.3 m/s for demonstration purposes. The combination of experimental tests and demonstration runs consumed approximately twelve hours of operation, although the actual time was not documented. The first attrition test evaluated a second batch of sodium sesquicarbonate over a twenty hour period at a gas velocity of 0.2 m/s. The two studies represented autogenous grinding condition for sodium sesquicarbonate in the FVU, and differences in physical properties for the two feed stocks, summarized in Table 1, were attributed primarily to void volume.

Observations of sample density and size confirmed that size reduction (attrition) depended on differences in particle density. Grinding produced an increased volume of fines over time: the amount of fine particles (between 74 and 126 micron) increased approximately 1.5 weight percent per hour, whereas the amount of course particles (between 178 and 417 micron) decreased approximately 0.9 weight percent per hour. Attrition rates, summarized in Table I for both feed stocks confirmed that particle density affected attrition of the media. Average particle size decreased 3.05 microns/h during the Hydrodynamic Investigation for a media which consisted of porous particles noted by the smaller bulk density. The media for the first attrition study consisted of dense particles and decreased 0.65 microns/h (average particle size). The porous particles abraded faster than the harder dense particles. Although the particle size distribution for the two feed stocks differed significantly, illustrated in Figure 1, grinding in the bed produced media with similar distributions, illustrated in Figure 2. These observation must be considered in future work and the final fluid-bed design; the remainder of this study, however, focused on the effects of velocity and time.

A gentle grinding action in the fluid bed abraded the surface which not only controlled the production of fines but also maintained a constant level of fines in the bed. Attrition rates based on average particle size, presented in Table 2, increased as the velocity increased and slightly as a grinding agent (catalyst base) was added to the bed. Analysis of the size distribution as a function of the Reynolds Number produced a power dependence with respect to particle diameter in the fines region (less than 200 micron), consistent with the hypothesis presented by Gaudin (4). The linear relationship, depicted in Figure 3, illustrated the distribution of fine material, of particular interest with respect to carry over from the bed, with minor changes over time. The fractions of fine material depended primarily on Reynolds Number, illustrated in Figure 4 and simulated by the correlation:

$$p(r) = 1.38 \left(\frac{D_p \langle v_g \rangle \rho_g}{\mu_g}\right)^{0.92}$$

$$= 1.38 N_{Re}^{0.92} \qquad 4$$

Closer examination of the results revealed that the distribution depended on a second variable, illustrated in Figure 5. A distinct shift to a coarser size distribution as the velocity increased implied that the distribution depended further on the velocity or in this case on the dimensionless time.

Wolff, Gerritsen and Verheijen (7) presented a model for attrition due to fluidization based on a combination of first and second order processes, which fit their data over the long term (10^4 h). They concluded that a first order process best described their data for the initial period (4 to 40 h). We analyzed our data based on an exponential relationship for time similar to their first order process. In general the fraction retained, illustrated in Figure 6, decreased exponentially with dimensionless time:

$$p(r) = 31.94 e^{(-2.87E-07 \frac{t \langle v_g \rangle}{D_p})}$$

$$= 31.94 e^{(-2.87E-07\theta)} \qquad 5$$

Although significant fluctuations in bed operations can occur during this initial operating period, Wolff et al. (7) noted that the average size will approach a constant value over the long term period.

We regressed the multivariable data using linear techniques to produce a correlation which described the distribution (weight fraction of material) of fines as a power function of Reynolds Number dependent exponentially on dimensionless time. The empirical equation:

$$p(r) = 3.39 N_{Re}^{0.92} e^{(-2.87E-07\theta)} \qquad 6$$

simulated the distribution within 25 percent, illustrated in Figure 7. Since both Reynolds Number and dimensionless time contained fluidizing velocity and particle diameter, the fraction retained varied as a power function of particle size, consistent with Gaudin (1939), exponentially with time, consistent with Wolff et al. (7). The grind coefficient (0.92) indicated that homogeneous mixing created by bubbling in the bed produced a gentle grind without producing an excess of fines. The rate coefficients (2.87E-07) confirmed the relatively slow production of fine particles over time.

CONCLUSION

This study improved our understanding of the attrition mechanisms associated with fluidization in a bubbling bed, proposed for the thermal-treatment process. Fluidizing velocity and media properties (diameter, density and void volume) determined the level of grinding and fines produced in the bed. Although particle size distribution for sodium sesquicarbonate feed stocks differed significantly, grinding due to fluidization produced bed media with similar distributions. Fluidizing velocity and particle size dominated as parameters which determined the characteristics of the bed. Size distribution depended primarily on fluidizing velocity (the major parameter for transport of mechanical energy to the bed and the comminution mechanism) and particle size. An empirical equation, which combined both velocity and particle size contributions, reasonably simulated sodium sesquicarbonate size distribution as a result of the grinding action in the bed.

Testing will continue to validate the relationship and to refine the model for the multicomponent media. We will interface this correlation with carry-over models, proposed by Kunii and Levenspiel (8), and will expand the model to include both grinding and carry-over for a bubbling bed unit.

NOTATION

C	constants
D_p	particle diameter (m)
F_2	carry over from bed (kg/s)
p(r)	particle size distribution (1/m)
R(r)	rate of particle size change (m/s)
N_{Re}	Reynolds Number
r	particle radius (m)
t	time (s)
$<v_g>$	gas velocity (m/s)
W	weight of the bed (kg)
μ_g	gas viscosity (kg/m-s)
ρ_g	gas density (kg/m³)
θ	dimensionless time

LITERATURE CITED

1. Mullen, J.F., *Chem. Eng. Prog.* **88**, 50, June 1992.

2. Frounfelker, R.E., *Solid Waste and Power*, **III**, 5, October 1989

3. Beke, B., Principles of Comminution, Publishing House of Hungarian Academy of Sciences, Budapest, 1964.

4. Gaudin, A.M., Principles of Mineral Dressing, McGraw-Hill Book Company, New York, New York, 1939.

5. Levenspiel, O., Kunii, D. and Fitzgerald, T. *Powder Technology*, **2**, 87-96, 1968.

6. Mauri, R. and Stecconi, P., *Sixteenth International Symposium on Combustion*, The Combustion Institute, 1981.

7. Wolff, E.H.P., Gerritsen, A.W., and Verheijen, P.J.T., *Powder Technology*, **76**, 47, 1993.

8 Kunii, D., and Levenspiel, O., Fluidization Engineering, Second Edition, Butterworth-Heinemann, Boston, 1991

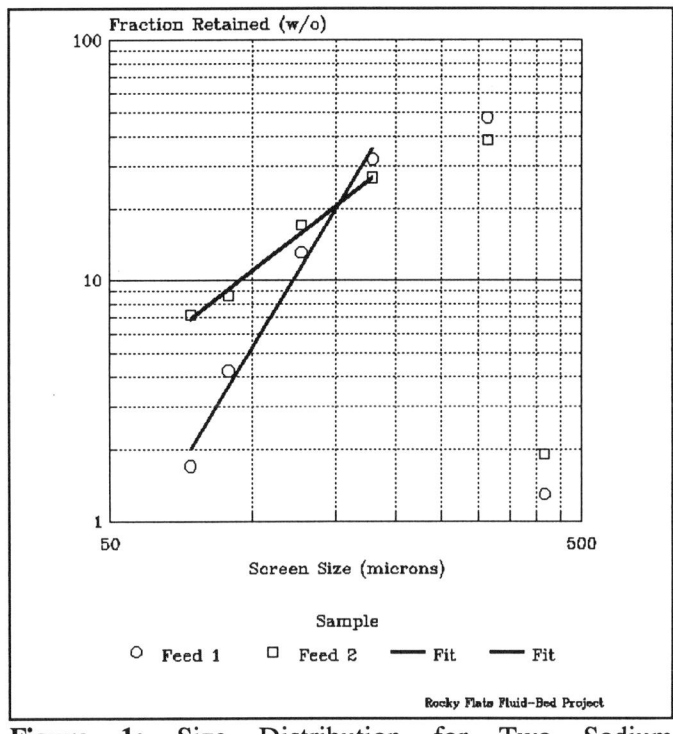

Figure 1: Size Distribution for Two Sodium Sesquicarbonate Feed Stocks.

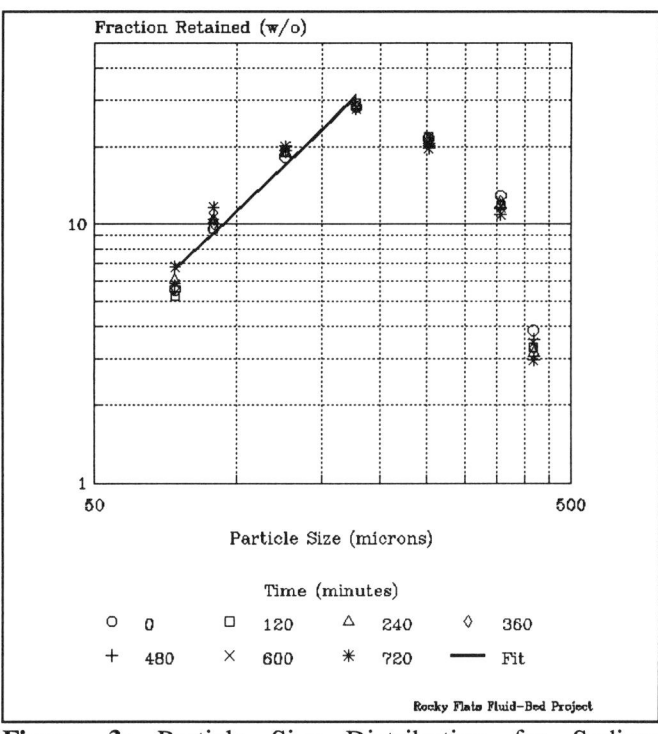

Figure 3: Particle Size Distribution for Sodium Sesquicarbonate over Time at a Velocity of 0.20 m/s.

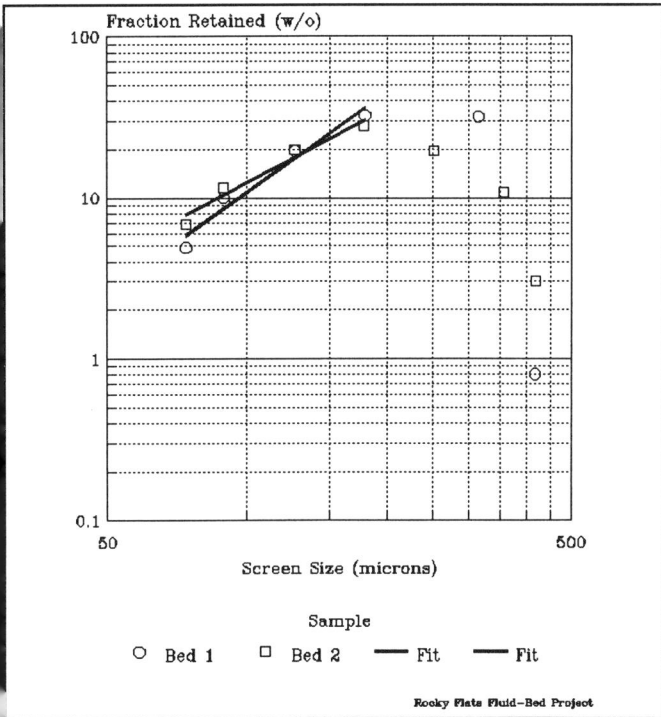

Figure 2: Size Distribution for Both Sodium Sesquicarbonate Beds after Several Hours of Operation.

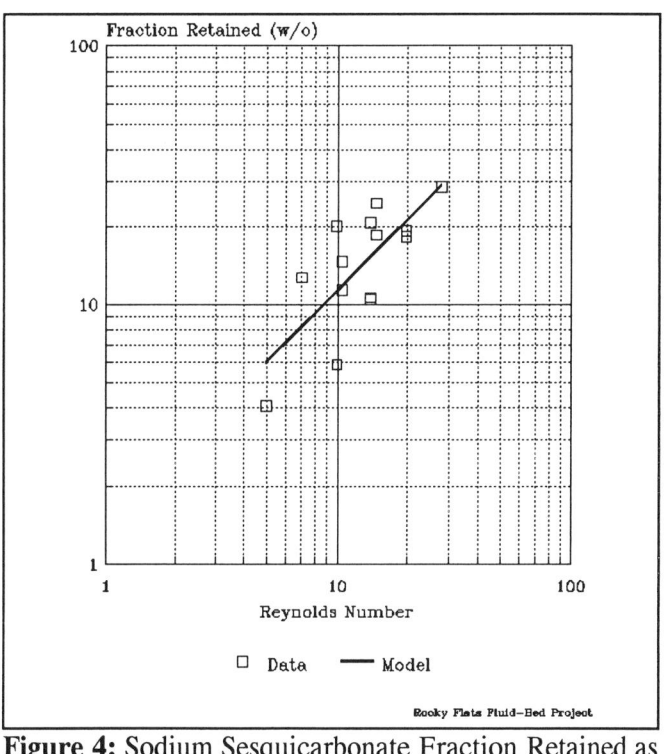

Figure 4: Sodium Sesquicarbonate Fraction Retained as a Function of Particle Reynolds Number.

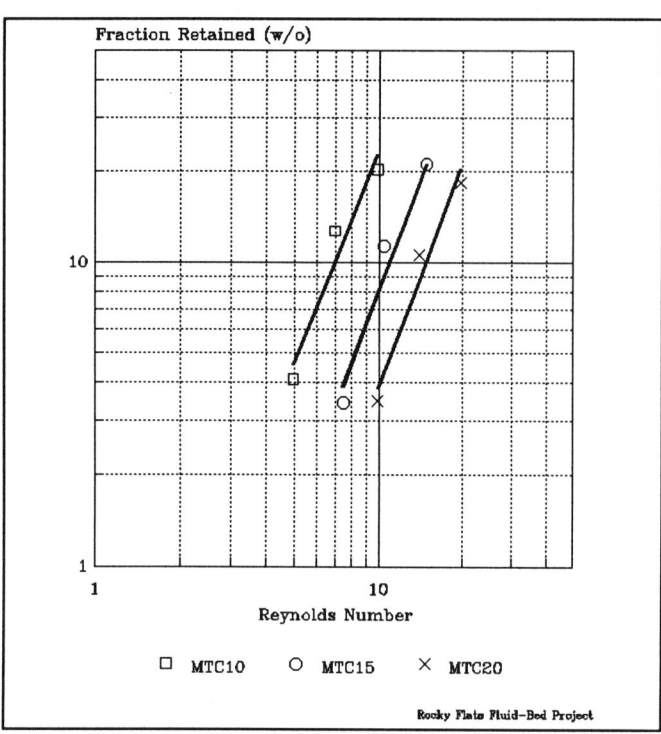

Figure 5: Reynolds Number Correlation Expanded to Illustrate an Additional Dependence due to Gas Velocity.

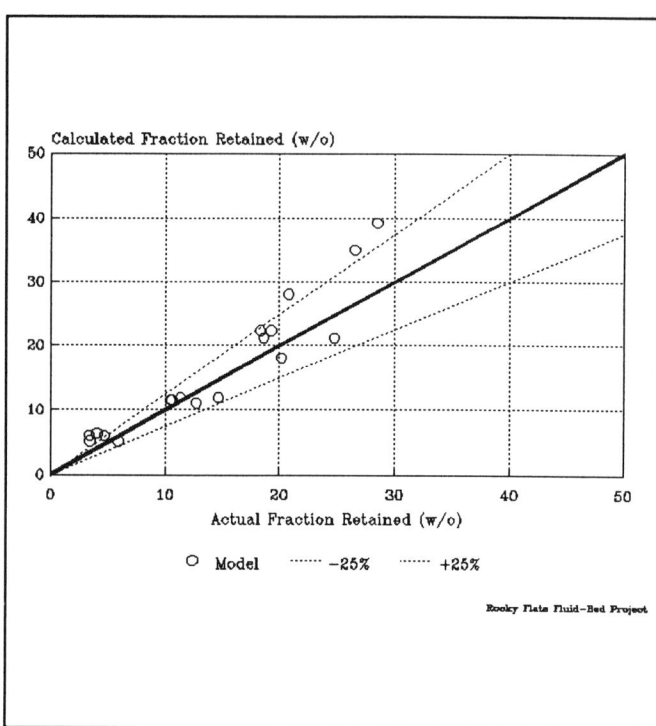

Figure 7: Verification of Attrition Model against Sodium Sesquicarbonate Size Distribution from the Flow Visualization Unit.

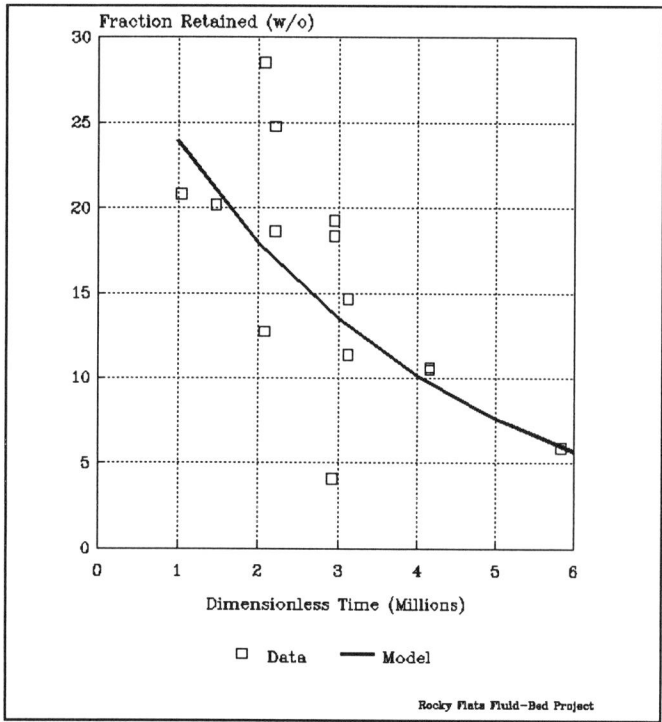

Figure 6: Sodium Sesquicarbonate Fraction Retained as a Function of Dimensionless Time.

Table 1
Summary of Properties for Sodium Sesquicarbonate Feed Samples Used in the Flow Visualization Unit
Rocky Flats Fluid-Bed Project

	Baseline	*Attrition*
Time (h)	12	20
Velocity (m/s)	variable	0.20
Density (kg/m^3) - feed - bed	 943 1088	 998 1030
Average Particle Size (microns) - feed - bed	 198 161	 191 172
Attrition (micron/h)	3.05	0.68

Table 2
Operating Conditions and Criteria Which Compare Attrition Results for Velocities of 0.15 and 0.20 m/s
Rocky Flats Fluid-Bed Project

Study	*Velocity (m/s)*	*Attrition (micron/h)*
Single Media		
SS15	0.15	0.451
SS20	0.20	0.667
Multi Media		
TC10	0.10	0.079
TC15	0.15	0.626
TC20	0.20	0.716

Electrostatic Spraying of Gases into Liquids

C. Tsouris, D.W. DePaoli, J.Q. Feng and T.C. Scott
Chemical Technology Division, Oak Ridge National Laboratory, P.O. Box 2008
Oak Ridge, TN 37831

In many chemical processes there is a need to disperse one fluid into another in the form of fine droplets or bubbles. In such operations, mechanical agitation is traditionally used, transferring the kinetic energy through the bulk fluids where a substantial amount is dissipated into thermal energy. Another technique to generate fine drops or bubbles of uniform size at the tip of a capillary tube is by using electric fields. Electrostatic dispersion can be more efficient because forces appear at interfaces without causing substantial electric current. Thus, this approach has found applications in many fields including manufacturing of ceramics, chemical processing, printing, spray painting, and crop spraying. Until recently, it was believed that electrostatic spraying may be used only in the case of dispersing a fluid into a relatively nonconductive fluid. Successful electrostatic dispersion of a nonconductive fluid into a conductive fluid has recently been reported in the literature. Experiments on electrostatic spraying of (i) gases into insulating and conducting liquids and (ii) liquids into gases are reported here. Measurements of bubble and drop size, pressure, and electric current during electrostatic spraying of gas-liquid systems including air-water, air-trichloroethylene, air-hexane, air-decane, nitrogen-ethanol, and nitrogen-water are obtained. Results show that electrostatic spraying of a gas into a liquid resembles the spraying behavior of a nonconductive fluid into a conductive fluid as described by Tsouris et al. (1994). As the conductivity of the surrounding fluid decreases, a higher applied voltage is needed for spraying.

Electrostatic spraying or dispersion is the phenomenon in which a fluid mass is disintegrated into numerous fine pieces by means of an electric field. Although physical effects of electric fields upon liquid surfaces were observed many years ago on surfaces of lakes, where waterspouts develop in the presence of thunderclouds, the first systematic study of electrified liquids was conducted by Abbe Nollet in the 18th century (Felici, 1959; Bailey, 1988). He observed that a wounded person does not bleed normally if he is electrified by connection to a high-voltage generator; instead, blood sprays from the wound.

Electrostatic spraying, according to Bailey (1984), is the result of the interaction between surface charge on a fluid meniscus and an externally applied field. The presence of the electric field drives charge carriers from the bulk of the dispersed liquid to the surface very rapidly. In the case of a capillary geometry, the force associated with the electric stress at the interface points outwards with respect to the meniscus. This description, however, refers only to electrostatic spraying of conductive fluids into nonconductive fluids, which, until recently, was believed to be the only successful spraying mode. Recent experiments by Sato and coworkers (1979, 1980a, 1980b, 1993) and Tsouris et al. (1994) showed that electrostatic spraying of nonconductive fluids into conductive fluids is possible when the capillary is adequately insulated.

Conducting experiments on electrostatic spraying of fluids, Tsouris et al. (1994) showed that, although there are differences in spraying nonconductive-in-conductive and conductive-in-nonconductive fluids, the two systems are consistent with the theory of electrohydrodynamics. It was found that, in the case of nonconductive-in-conductive, the force associated with the electric stress acts inwards with respect to the meniscus, whereas in the case of conductive in-nonconductive, the force acts outwards. This behavior was confirmed by pressure measurements inside the capillary as well as by the shape of the meniscus. More details of the spraying mechanism of insulating fluids are discussed by Feng et al. (1995). Experiments were then conducted to show the effects of various parameters such as distance between the capillary electrode and the ground electrode; the distance between the metal tip and the ceramic-insulation tip of the capillary; the wall thickness of the ceramic insulation tube; the flow rate of the dispersed fluid; the conductivity of the continuous phase liquid; the applied voltage; and the electrical current on the spraying behavior (Tsouris et al., 1995). Most of these experiments were conducted with liquid-liquid systems.

Electrostatic spraying of nonconductive fluids into conductive fluids may have applications in a plethora of industrial processes. Among these applications are processes in which a gas is dispersed into a liquid. Although some gases are more conductive than some liquids, the majority of gases are less conductive than the majority of liquids. Therefore spraying a gas into a liquid generally falls in the category of a nonconductive fluid sprayed into a more conductive fluid. Electrostatic spraying of a gas into a liquid is an efficient method of generating fine bubbles of relatively uniform size. This method is proposed here to replace such devices as gas spargers or diffusers (Figure 1) that are traditionally used to introduce a gas into a liquid, as well as mechanical agitation, that is used to break up the ejected bubbles into smaller size. The reason that agitation is less effective than electrostatic spraying is that it transfers the kinetic energy through the bulk fluids, where a substantial amount is dissipated into thermal energy. In addition, only limited control of the bubble size distribution is provided by agitation.

The objective of this article is to discuss electrostatic spraying of gases into liquids as an efficient method of creating fine gas-liquid dispersions. The experimental system will first be discussed, followed by the presentation of experimental findings and discussion of the results.

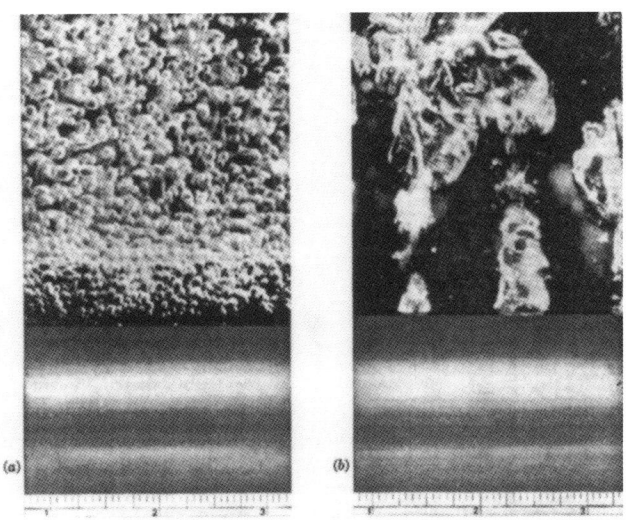

Figure 1. (a) Porous-carbon diffuser; (b) Perforated-pipe sparger (R. H. Perry and D. Green, Perry's Chemical Engineers' Handbook, 6th Edition, 1984, p. 18-62, McGraw-Hill. Reproduced with permission of McGraw-Hill, Inc.)

Materials and Methods

Experimental Assembly. The effects of an electric field on spraying behavior may be examined by an experimental arrangement shown in Figure 2. A metal capillary, electrically connected to a high-voltage power supply, is used for spraying. The capillary is constructed of electropolished and electrolytically cut capillary tubing obtained from Valco Instruments (tubing kits T10N5D, T10N10D, and T10N15D). The metal capillary is electrically insulated from the surrounding liquid by a ceramic or glass tube. Different inside and outside diameters for both the capillary electrode and the ceramic insulation tube have been tested successfully; however, it has been observed that electrostatic spraying occurs at a lower applied voltage, with smaller-diameter capillaries. Experiments were conducted with different dimensions of the stainless steel capillary electrode, varying in ID from 0.125 mm to 0.5 mm and in OD from 0.8 mm to 1.6 mm, and for the ceramic-insulation tube, varying in ID from 0.8 mm to 1.6 mm and in OD from 1.6 mm to 3.2 mm. No additional insulation or sealant was applied between the capillary electrode and the ceramic insulation, and, therefore, the distance between the tip of the metal and the tip of the ceramic could be easily adjusted. The outside diameter of the ceramic insulation was used for calibration of drop or bubble size measurements. The gas is driven through the capillary inside a rectangular glass tank contactor of dimensions 50 mm X 50 mm X 200 mm by a syringe pump (Razel, model A-99). The voltage signal from the high-voltage power supply can be AC, DC, or pulsed DC. In the experiments described here, DC (Bertan Associates, Inc., Series 225) and pulsed DC (built in our laboratories) power supplies were used. The DC supply operates in a current range of 0-0.3 mA and provides digital readings of the applied voltage and electrical current. The metal capillary is connected to the power supply output, thus serving as the electrified electrode, while the second electrode is a 3-mm-diameter metal rod located above the end of the capillary at an adjustable spacing. This second electrode is grounded. Visualization equipment including a high-speed imaging system (Kodak, Ektapro Intensified Imager; maximum speed 12,000 frames per second), a video recorder (Panasonic, AG 1960 proline), a monitor (Audiotronics, 14VM939), and a printer (Mitsubishi, Video copy processor) are used to visualize and print images of drops generated at the end of the metal capillary. Also, shown in Figure 2 is a sensitive pressure transducer (ENDEVCO Σ, Model 8510B-1) operating in a pressure range of 0-1 psig and connected to a data acquisition system. This transducer is used to monitor pressure variations in the capillary.

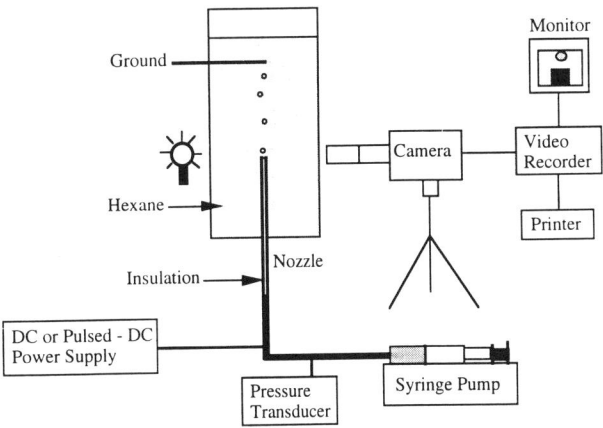

Figure 2. Experimental assembly for electrostatic spraying of gases.

Table 1. Physical Properties (20 °C)

Substance	Density Difference ($\rho^* - \rho_{water}$) (g/cm³)	Viscosity (cP)	Interfacial Tension with Water (dyn/cm)	Conductivity (µmho/cm) (µS/cm)
Water	-----	1.00	-----	0.3—13
Decane	-0.27	0.93	25.7	not available
Hexane	-0.34	0.31	37.5	10^{-12}
TCE	0.46	0.60	37.0	10^{-2}
Air	-1.00	0.002	72.0	-----
Ethanol	-0.19	1.40	(-air) 22.3	1.5
Nitrogen	-1.00	0.002	-----	-----

* Substance other than water.

Materials. The sample fluids that have been used in this study are air - distilled water, air - trichloroethylene (TCE, Fisher), air - hexane isomers (Fisher), air - decane, nitrogen - distilled water, nitrogen - ethanol (95% ethanol - 5% water), and nitrogen - TCE. The physical properties of these fluids are given in Table 1. The interfacial tension and conductivity measurements of water were obtained in our laboratory using a Fisher Autotensiomat and a conductivity meter from The London Company (CDM 2e). The conductivity of TCE was measured in the laboratories of Scientifica (Conductivity Meter, Model 627), manufacturer of low-conductivity meters. All the other properties were obtained from the literature (e.g., <u>CRC Handbook of Chemistry and Physics</u>). The flow rate of the dispersed fluid was maintained constant for the duration of each experiment.

Results

In this study, we report experiments on electrostatic spraying of gas-liquid systems. Bubble and drop size, current, and pressure measurements are presented versus applied voltage. These results demonstrate the effectiveness of electric fields in creating large surface areas for mass transfer in gas-liquid systems.

The effect of applied DC voltage on the bubble size of air in decane is shown in Figure 3. In this experiment, the voltage was varied from 0V to 22.5kV. The bubble size is shown to decrease with increasing applied voltage. A pinch-off mode of bubble formation is observed in Figure 3e (17.5kV). As discussed by Tsouris et al. (1994), this mode of bubble or drop formation is observed when spraying a relatively nonconductive fluid into a conductive fluid. In the inverse mode of electrostatic spraying, i.e., spraying a conductive fluid into a nonconductive fluid, a conical drop which jets from the tip is observed. In the case of air sprayed into decane, although decane is a very nonconductive liquid, the behavior of a nonconductive fluid sprayed in a conductive fluid is observed because air is less conductive than decane. A similar behavior has been observed for the systems air-in-water, air-in-TCE, air-in-hexane, nitrogen-in-water, and nitrogen-in-TCE.

The air-in-decane experiment with DC applied voltage produces uniform bubble size as shown in Figure 3. In the case of a pulsed DC applied voltage, however, a very broad size distribution may be produced at a high pulsing frequency (see Figure 4).

Drop and bubble size results for the TCE-in-air and air-in-TCE systems are presented in Figure 5. For the TCE-in-air system, the distance between the end of the metal capillary and the ground electrode is 6cm, whereas, for the air-in-TCE system, this spacing was varied from 4cm to 8cm. A 0.8-mm-OD capillary without insulation was used in this experiment. The TCE-in-air system is shown to have a stronger response than air-in-TCE to variations in DC applied voltage. Jetting of TCE, which produces very fine drops, occurs at 3.75kV. This jetting is a result of an outward force due to the electric stress at the fluid interface. The case of air-in-TCE is essentially a nonconductive-in-conductive case in which, as discussed by Tsouris et al. (1994), the electric stress results in an inward force. Apparently, this force, even at as high as 20kV, could not create a jetting mode of spraying, suggesting that it is relatively easier to spray a liquid than a

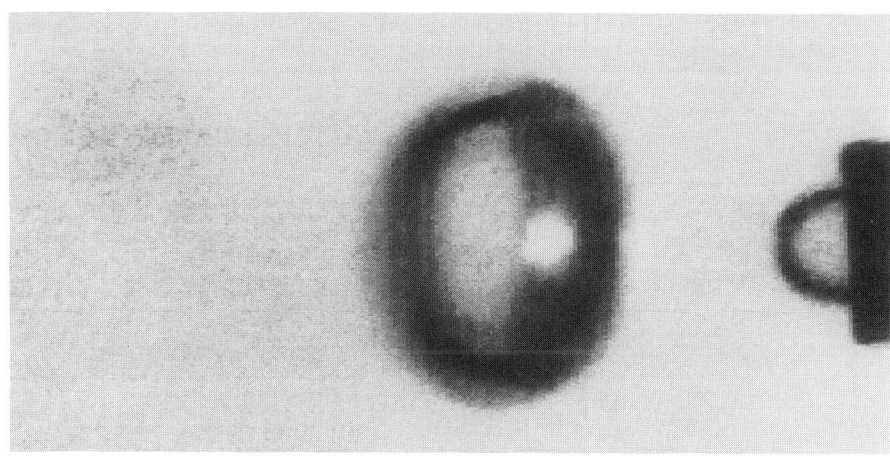

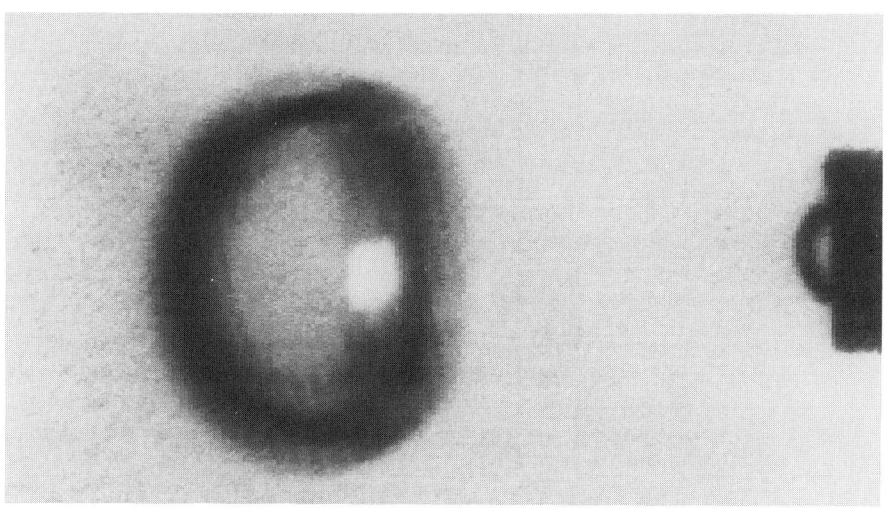

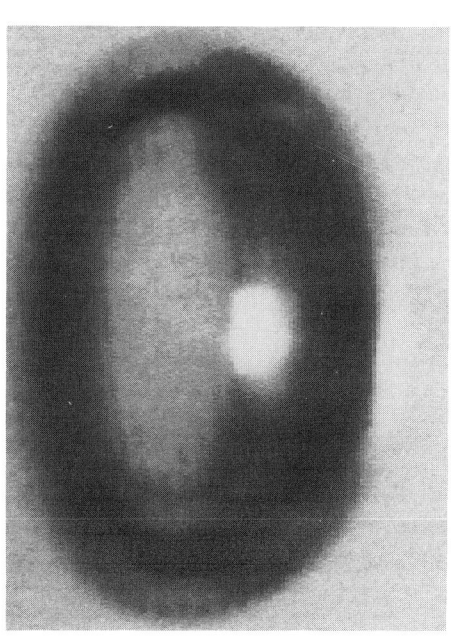

Figure 3. Spraying of air into decane.

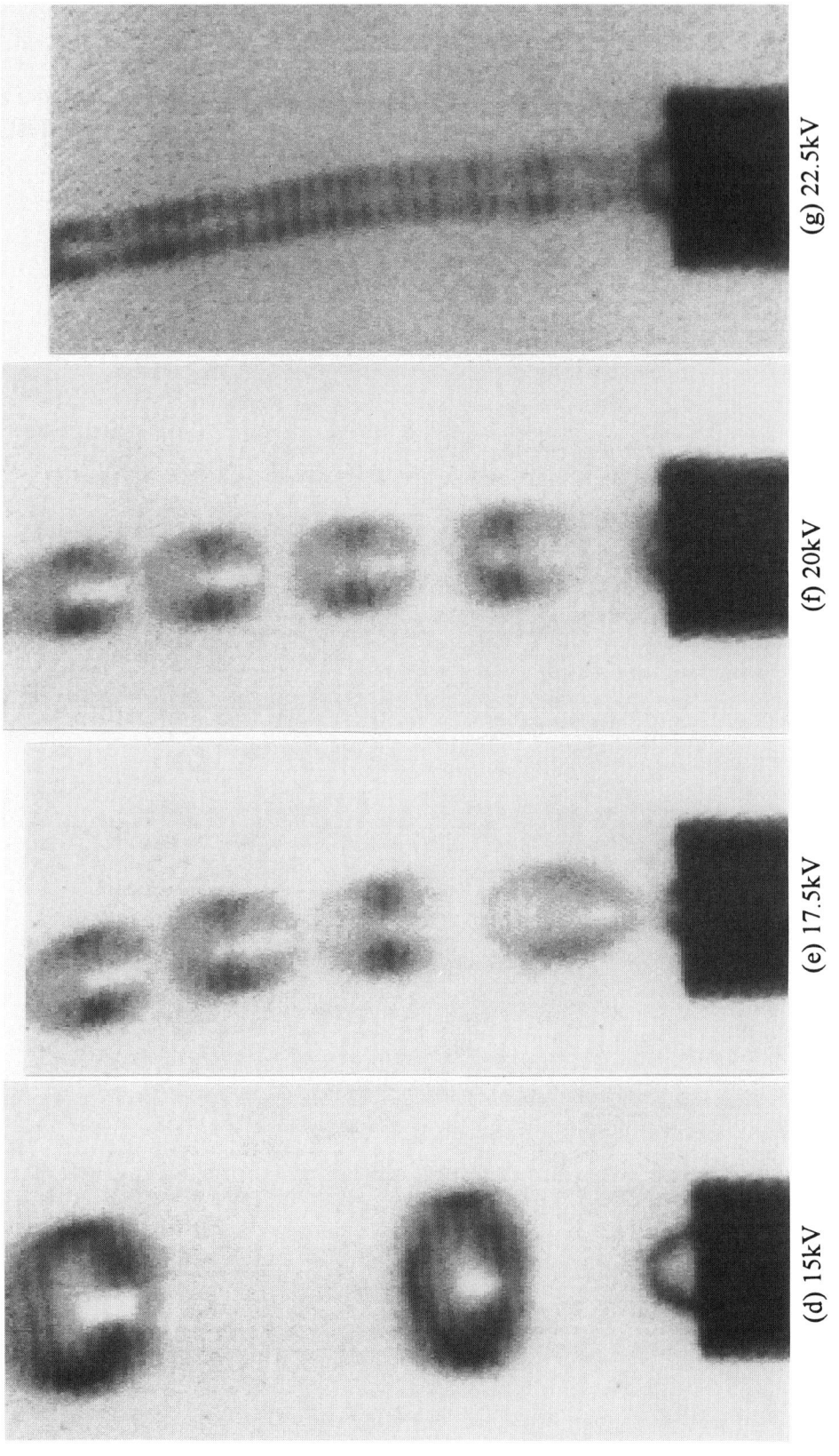

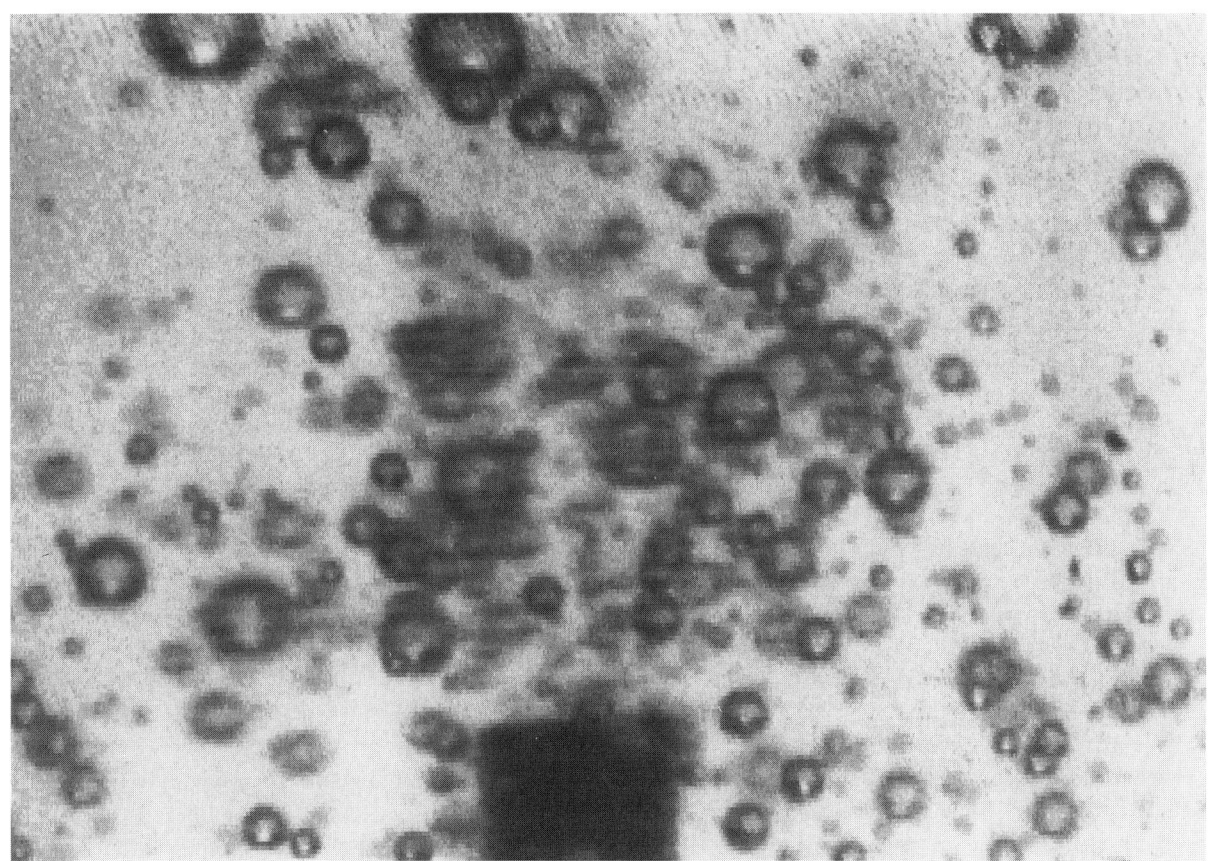

Figure 4. Spraying of air into decane by a high-frequency pulsed DC applied voltage.

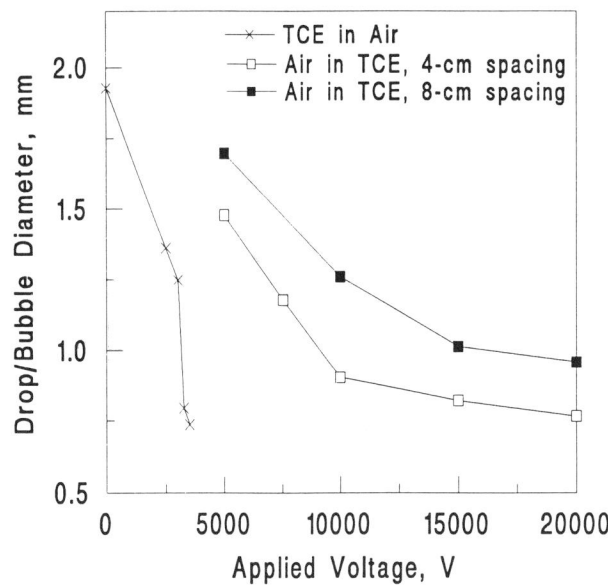

Figure 5. Effect of applied DC voltage on drop/bubble size.

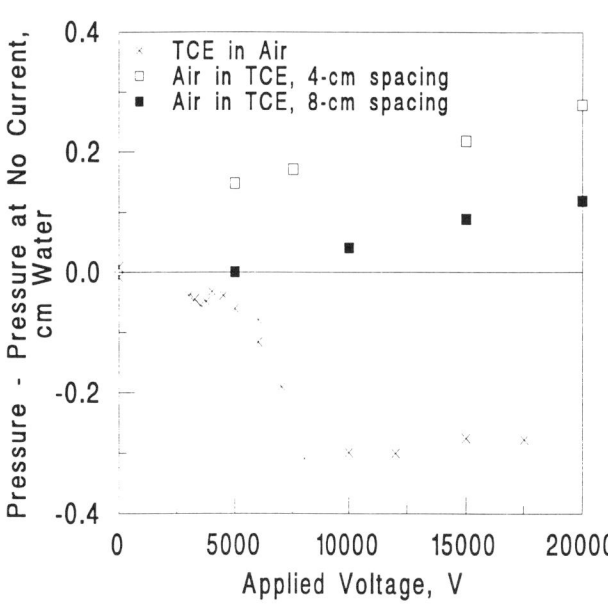

Figure 6. Effect of applied DC voltage on pressure inside the capillary.

gas. Although the air-in-TCE behaves as a nonconductive-in-conductive system with respect to the force direction, the spacing between the metal capillary and the ground has a significant effect on bubble size. The reason is that TCE, although a better electrical conductor than air, is not a good conductor in general.

The direction of the electric force is verified by pressure measurements which are presented in Figure 6. The pressure inside the capillary increases with applied voltage for the case of air-in-TCE and decreases for the case of TCE-in-air. These measurements suggest that the force due to the electric stress acts inwards for the air-in-TCE system and outwards for the TCE-in-air system. These systems are described by Tsouris et al. (1994) as nonconductive-in-conductive (air-in-TCE) and conductive-in-nonconductive (TCE-in-air).

Measurements of current for the same experiments presented in Figs. 5 and 6 are shown in Figure 7. For the air-in-TCE system, the expected result of lower current, when the spacing between the capillary and the ground is shorter, is observed. For the case of TCE-in-air, a significant increase of current is observed with increasing applied voltage above 10kV. Above this voltage, multiple jets were formed at the tip of the metal capillary. It is speculated that corona discharge occurred at these conditions. This speculation is supported by the flat Pressure-curve for TCE-in-air above 10kV in Figure 6 and the change in the slope of the Current-curve in Figure 7.

All gas-liquid systems studied in this work, i.e., air-water, air-TCE, air-decane, air-hexane, nitrogen-water, and nitrogen-TCE, showed a spraying behavior similar to the air-TCE and air-decane systems that have been discussed above. Experimental results show that electric fields can effectively be used to create large amounts of surface area for mass transfer between gases and liquids. On-going research is focused on quantifying the effect of several parameters on the spraying behavior. The completion of this study will provide answers on the effectiveness and applicability of electrostatic spraying in gas-liquid chemical processing.

One of the many important parameters of electrostatic spraying of gases into liquids is the pulsing frequency of pulsed DC electric fields. At a sufficient intensity of pulsed-DC applied voltage and a pulsing frequency of approximately 1000 Hz, for example, a very small bubble size of air-in-decane was produced. This bubble size is compared to non-electric-field conditions in Figure 8. The size of the bubbles was found to be very sensitive to frequency variations. Experiments with finer frequency control are needed to better investigate this phenomenon. Another example of the importance of pulsing frequency is shown in Figure 9. In this experiment, a dark cloud of apparently fine bubbles is observed. The results of both Figs. 8 and 9 indicate that, at certain conditions, it is possible to form a very fine dispersion of a gas into a liquid with small energy consumption.

Discussion

The work presented here demonstrates electrostatic spraying of gases in liquids and liquids in gases. Spraying of a gas into a liquid follows the behavior of a nonconductive fluid sprayed into a conductive fluid as discussed by Tsouris et al. (1994). For the conductive-in-nonconductive system (e.g., water in air), the drop at the capillary is elongated and very fine drops are released from its tip toward the second electrode. This behavior is a result of a strong outward force that appears at the interface (see also Sample and Bollini, 1972). At higher applied voltages, jetting occurs in various directions as the tip of the attached drop shifts alternately from one side of the capillary to the other (Zeleny, 1915; Cloupeau and Prunet-Foch, 1989). In the inverse case, electrical forces are directed inward from the conductive to the nonconductive fluid and act at the tip of the capillary, forcing the release of fine drops or bubbles. The higher the electric field, the stronger the inward forces become and the smaller the drops that are generated. Inward forces are stronger near the wall of the metal capillary, where the thickness of the attached nonconductive drop or bubble is minimum; therefore, drop release occurs at the center of

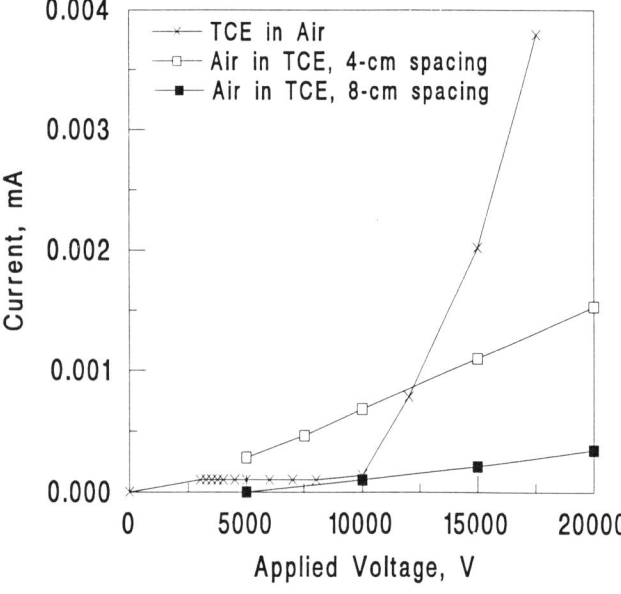

Figure 7. Effect of applied DC voltage on electrical current.

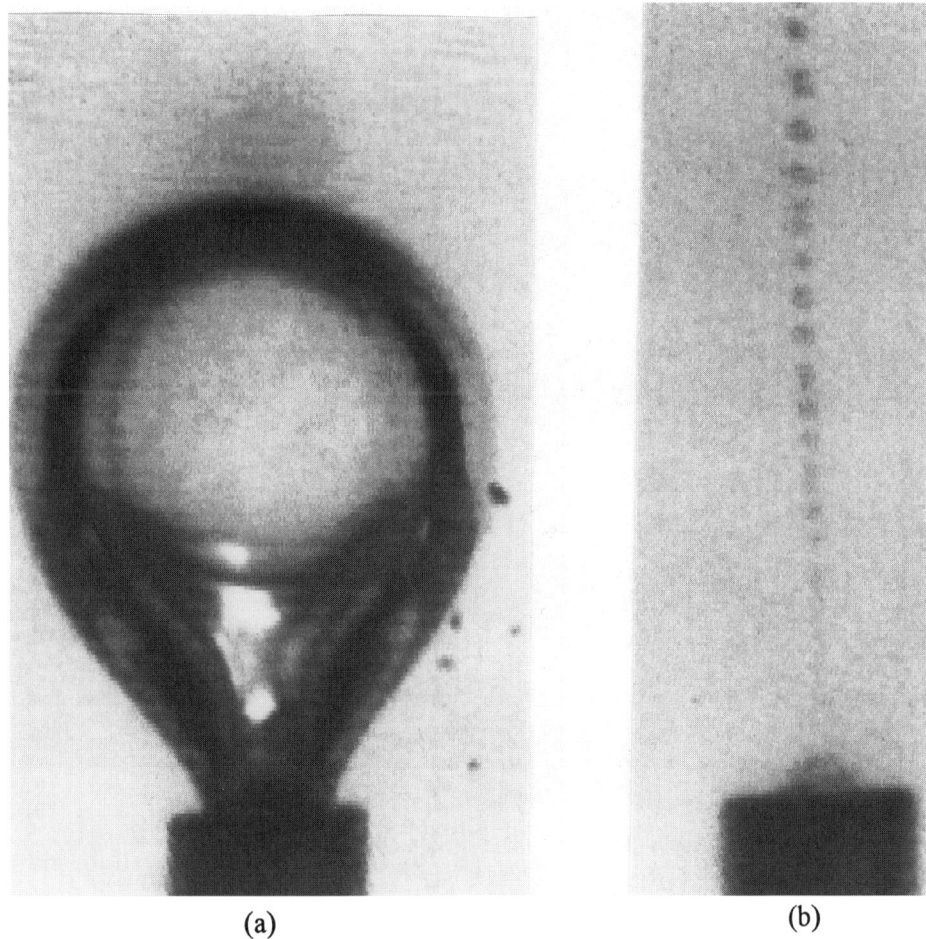

Figure 8. Bubble size produced by
(a) no voltage applied; (b) pulsed DC voltage applied.

the capillary, right at the tip, as a result of a pinch-off action.

Spraying a gas into a conductive liquid is only possible by appropriately insulating the metal capillary as shown in Figure 3. As the conductivity of the surrounding phase increases, however, special care of the insulation of the capillary must be taken, otherwise dielectric breakdown or sparking occurs (Tsouris et al., 1995). Currently, work is focused in spraying gases into highly conductive liquids and in correlating the size of the emitted drops or bubbles with the pertinent parameters of the system. For the latter, a dimensional analysis approach utilizing a sufficient number of experimental data from various chemical systems and geometric configurations is considered.

Summary

In summary, the experimental results presented here demonstrate electrostatic spraying of gases into liquids as a new technique to create large amounts of surface area for mass transfer and chemical reactions in gas-liquid systems. In earlier studies on liquid-liquid dispersions (Scott, 1989; Scott and Wham, 1989; and Scott et al., 1994), it has been shown that using electric fields to create emulsions is a much more efficient method than frequently used techniques. It is anticipated that electrostatic spraying of gases into liquids will be as effective as spraying conductive liquids into nonconductive liquids. If this is the case, this new technique for spraying gases into liquids may find applications in such areas as environmental technology, bioprocessing, and chemical industry.

Acknowledgments

Funding provided by the Division of Chemical Sciences, U.S. Department of Energy, under contract DE-AC05-84OR21400 with Martin Marietta Energy Systems, Inc., is gratefully acknowledged.

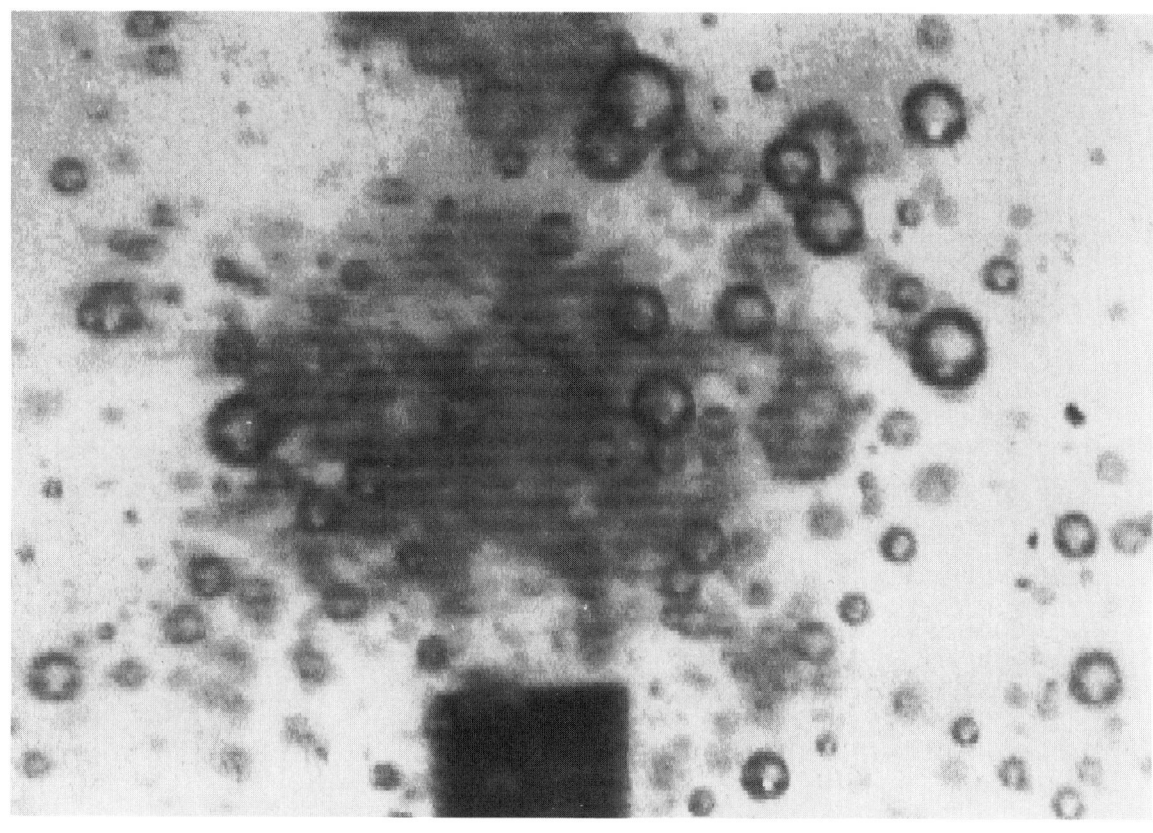

Figure 9. Fine bubble size produced by pulsed DC high-frequency, high-voltage signal.

References

Bailey, A. G., "Electrostatic Spraying of Liquids", *Phys. Bull.*, **35**, 146 (1984).

Bailey, A. G., "Electrostatic Spraying of Liquids". Research Studies Press Ltd: Taunton, Somerset, England (1988).

Cloupeau, M., and B. Prunet-Foch, "Electrostatic Spraying of Liquids in Cone-Jet Mode", *J. Electrostat.*, **22**, 135 (1989).

Felici, N. J., "Recent Developments and Future Trends in Electrostatic Generation", *Direct Curr.*, **4**, 192 (1959).

Feng, J. Q., D. W. DePaoli, C. Tsouris, and T. C. Scott, "Spraying Fine Fluid Particles in Insulating Fluid Systems by Electrostatic Polarization Forces, submitted in *Journal of Applied Physics* (1995).

Perry, R. H., and D. Green, "Perry's Chemical Engineers' Handbook", Sixth Edition, McGraw-Hill, New York, p. 18-62 (1984).

Sample, S. B., and R. Bollini, "Production of Liquid Aerosols by Harmonic Electrical Spraying", *J. Colloid Interf. Sci.*, **41**, 185 (1972).

Sato, M. J., "Cloudy Bubble Formation in a Strong Nonuniform Electric Field", *Electrostat.*, **8**, 285 (1980a).

Sato, M., M. Kuroda, and T. Sakai, "Effect of Electrostatics on Bubble Formation", *Kagaku Kogaku Ronbunshu*, **5**, 380 (1979).

Sato, M., M. Saito, and T. Hatori, "Emulsification and Size Control of Insulating and/or Viscous Liquids in Liquid-Liquid Systems by Electrostatic Dispersion", *J. Colloid Interf. Sci.*, **156**, 504 (1993).

Sato, M., Y. Takano, M. Kuroda, and T. Sakai, "A New Cloudy-Bubble Tracer Generated under Electrostatic Field", *J. Chem. Eng. Jpn*, **13**, 326 (1980b).

Scott, T. C., "Use of Electric Fields in Solvent Extraction: A Review and Prospectus", *Sep. Purif. Meth.*, **18**, 65 (1989).

Scott, T. C., and R. M. Wham, "An Electrically Driven Multistage Countercurrent Solvent Extraction Device: The Emulsion-Phase Contactor", *Ind. Eng. Chem. Res.*, **28**, 94 (1989).

Scott, T. C., D. W. DePaoli, and W. G. Sisson, "Further Development of the Electrically Driven Emulsion-PhaseContactor", *Ind. Eng. Chem. Res.*, **33**, 1237 (1994).

Tsouris, C., D. W. DePaoli, J. Q. Feng, O. A. Basaran, and T. C. Scott, "Electrostatic Spraying of Nonconductive Fluids into Conductive Fluids", *AIChE J.*, **40**, 1920 (1994).

Tsouris, C., D. W. DePaoli, J. Q. Feng, and T. C. Scott, "Experimental Investigation of Electrostatic Dispersion of Nonconductive Fluids into Conductive Fluids", *Ind. Eng. Chem. Res.*, in press (1995).

Zeleny, J., "On the Conditions of Instability of Electrified Drops, with Applications to the Electrical Discharge from Liquid Points", *Proc. Cambridge Philos. Soc.*, **18**, 71 (1915).

Mass Transfer in a Laminar Rippling Film in a Conical Centrifugal Film Reactor

Part I: Film Thickness Profile and Velocity Profile

Richard Long and Tushar Pattni
New Mexico State University, Department of Chemical Engineering, Las Cruces, NM 88003

The film thickness profile and the velocity profile for film flow on a rotating cone are computed in this section. These profiles are necessary in part II, and part III in order to evaluate stability and mass transfer.

INTRODUCTION

The purpose of this work was to determine theoretically, the mass transfer coefficient for the transport of oxygen from the gaseous phase to the liquid phase in the region of laminar flow with rippling. The apparatus and the theory are discussed in Long et al's publication [1]. The reactor, called the Centrifugal Film Reactor (CFR), uses rotational motion to create a thin film which flows outward from the frustum of the cone. The flow of this film is in the regime of laminar flow with rippling [1]. Alfa-Laval constructed fast rotating centrifuges which consisted of conical discs closely spaced, with spacers being utilized for structural reasons. Without the spacers, as is the present case, the flow was confined to essentially thin Ekman layers [5].

The framework followed to accomplish this analysis was as follows:

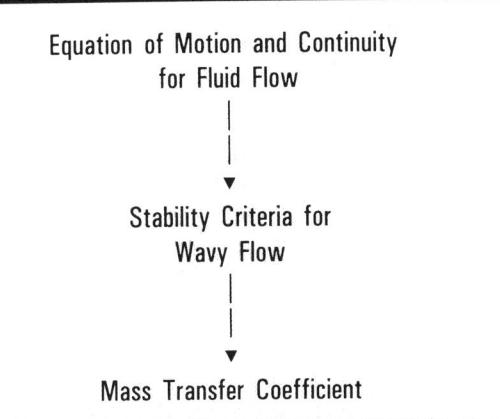

Equation of Motion and Continuity for Fluid Flow
↓
Stability Criteria for Wavy Flow
↓
Mass Transfer Coefficient

Dept. of Chemical Engineering, New Mexico State University, Las Cruces, New Mexico, Dr. Richard Long & Tushar Pattni

The equations of motion in this case are none other than the Navier-Stokes equations, since the fluid has constant density and viscosity. Therefore it is a momentum balance equation. It is mentioned by Greenspan [2] that there are some primary phenomena brought about by rotation. One is the tendency for the fluid motion to become two dimensional when the Rossby number to be low (neglecting viscous effects). This is known as the Taylor-Proudman theorem. The Rossby number expresses the relative importance of inertia and Coriolis force, i.e $\epsilon = U/Lf$. U is the fluid velocity relative to the rotating frame, L is the horizontal length scale and $f = 2(\Omega)$ (the Coriolis frequency). Inertial waves are another manifestation of rotation effects [6]. They exist when the fluid element oscillates at a frequency which is less then the Coriolis frequency. For larger frequencies, that is, when it is greater than the Coriolis frequency, the flow field created is the same as in non-rotating fluids. Various investigators [8,9,10] have found that the presence of wavy flow has the effect of increasing the mass and heat transfer, with the mass transfer rates being increased by about 200 %. Since ripples had been observed during the operation of the CFR, the purpose of this theoretical analysis was to determine if there was a significant increase in the mass transfer coefficient due to these ripples. Additionally, the contribution to the overall mass transfer, of the region under investigation was also of interest. This was due to the fact that there were five regions in which the mass transfer was taking place (Figure 1). The region under investigation in this study is Region 2.

FLUID FLOW ANALYSIS

Velocity Profiles

The cone rotates at angular velocity Ω radians/second, (Figure 2), and the horizontal plane from which the film rises inside the cone is at a radius Ro (Figure 3).

<u>Assumptions:</u> (1) The film thickness δo is much smaller than

the radius R_o, i.e $\delta o/R_o \leq 1$.

(2) Flow within the layer is viscous.

(3) Let the coordinate system rotate at an angular velocity Ω, therefore $v'\phi = -v'\phi + \Omega r \sin\Theta$, where $v\phi$ = tangential velocity.

The governing equations are the momentum and mass balance equations, i.e,

$$\frac{\partial \vec{v}}{\partial t} + \vec{v}\cdot\vec{\nabla}\vec{v} + 2\vec{\Omega}\vec{k}\times\vec{v} = \frac{-\vec{\nabla}p}{\rho} + g\vec{k}\rho + \nu\vec{\nabla}^2\vec{v} \quad (a)$$

and

$$\vec{\nabla}\cdot\vec{v} = 0 \quad (b)$$

F-1

Here p is the reduced pressure, given by the following expression,

$$p = P - \frac{1}{2}\rho(\vec{\Omega}\times\vec{r})\cdot(\vec{\Omega}\times\vec{r})$$

With the given assumptions, the momentum balance and the continuity equations were reduced to in component form (in the spherical coordinate system):

F-2

$$v_r\frac{\partial v_r}{\partial r} + \frac{v_\theta}{r}\frac{\partial v_r}{\partial \Theta} - \frac{v'^2_\phi}{r} - \Omega^2\sin^2\Theta$$

$$+2\Omega \acute{v}_\phi \sin\Theta = g\cos\Theta - \frac{1}{\rho}\frac{\partial p}{\partial r} + \frac{\nu}{r^2}\frac{\partial v_r}{\partial \Theta^2}$$

F-3

$$-v'^2_\phi\frac{\cot\theta}{r} - \Omega^2 r\sin\theta\cos\theta + 2\Omega v'_\phi\cos\theta$$

$$= g\sin\Theta - \frac{1}{\rho r}\frac{\partial p}{\partial \Theta} + \frac{\nu}{r}(\frac{\partial^2 v_\theta}{\partial \Theta^2} + 2\frac{\partial v_r}{\partial \Theta})$$

F-4

$$-v_r\frac{\partial v_\phi}{\partial r} - \frac{v_\theta}{r}\frac{\partial v'_\phi}{\partial \Theta} - \frac{v_r v'_\phi}{r} + 2\Omega v_r \sin\Theta$$

$$= -\frac{\nu}{r^2}\frac{\partial^2 \acute{v}_\phi}{\partial \Theta^2}$$

and

F-5

$$\frac{\partial v_r}{\partial r} + 2\frac{v_r}{r} + \frac{1}{r}\frac{\partial v_\theta}{\partial \Theta} = 0$$

The following further simplifications were then made:
- The pressure and the viscous terms (Equation (F-1) are one order of magnitude smaller than the corresponding terms, and so they are dropped.
- Replace Θ by angle ß, where the cone has angle 2ß.

Introducing the following dimensionless groups:

F-6

$$-ds = rd\theta \qquad \eta = \frac{r}{R_o} \qquad \sigma = (\frac{\Omega}{\nu})^{1/2}$$

F-7

$$Fr = \frac{\Omega^2 r \sin\beta}{g} \qquad Q_o^+ = \frac{Q}{2\pi \sin^2\beta r^2 (\Omega\nu)^{1/2}}$$

F-8

$$\Omega^+ = \frac{\Omega \sin\beta R_o}{\langle v_r|_{r=R_o}\rangle} \qquad \delta^+ = (\frac{\Omega}{\nu})^{1/2}\delta$$

$$U = \frac{V_r}{\Omega r \sin\beta} \quad V = \frac{V_\theta}{(\Omega \nu)^{1/2}} \quad W = \frac{V'_\phi}{\Omega r \sin\beta} \qquad \text{F-9}$$

$$P = \frac{p - p_o}{\rho (\Omega \nu)^{1/2} \Omega r \sin\beta} \qquad \text{F-10}$$

where,
    s is the coordinate introduced by equation F-6
    σ, η are dimensionless coordinates in s and r directions
    Fr is the Froude number
    Q_o^+ is the dimensionless volumetric flow rate
    Ω^+ is the dimensionless angular velocity
    δ^+ is the dimensionless film thickness
    U, V, W are the dimensionless velocity in r, Θ and ϕ directions
    P is the dimensionless pressure

After introducing the average radial velocity $<v_r|_{r=R_0}>$, to make v_r, v_θ and v_ϕ dimensionless,

$$U_0 = \frac{v_r}{<v_r|_0 r = R_0>} = \Omega^+ \eta U$$

$$V_o = \frac{v_\theta}{<v_r|_{r=R_o}>} = \frac{\Omega^+}{R_o \sin\beta}\left(\frac{\nu}{\Omega}\right)^{1/2} V$$

$$W_o = \frac{v_\phi}{<v_r|_{r=R_o}>} = \Omega^+ \eta W \qquad \text{F-11}$$

and introducing together with the dimensionless parameters, into the equations of motion and continuity, the following equations are obtained:

$$U_o^2 + \eta^2 U_o \left(\frac{U_o}{\eta}\right)_\eta - \frac{\eta V_o}{\Gamma \sin\beta}(U_o)_\sigma - W_o^2 \qquad \text{F-12}$$

$$-\Omega^2 \eta^2 + 2\Omega\eta W_0 = -\Omega^{+2}\eta^2 \frac{\cot\beta}{Fr} + \Omega^+ \eta \frac{\partial^2 U_o}{\partial \sigma^2}$$

$$2U_o W_o + \eta U_o \left(\frac{W_o}{\eta}\right)_\eta - \frac{\eta V_o}{\Gamma \sin\beta}\frac{\partial W_o}{\partial \sigma} + 2\eta\Omega U_o \qquad \text{F-13}$$

$$= \Omega \eta \frac{\partial^2 W_o}{\partial \sigma^2}$$

$$-W_o^2 - \Omega^2 \eta^2 + 2\Omega\eta W_o = \Omega^2 \eta^2 \frac{\tan\beta}{Fr} \qquad \text{F-14}$$

$$+ \Omega^2 \eta^2 \tan B \frac{\partial P}{\partial \sigma}$$

and

$$\eta\left(\frac{U_o}{\eta}\right)_\eta + 3\frac{U_o}{\eta} - \frac{\Gamma}{\sin\beta}\frac{\partial V_o}{\partial \sigma} = 0 \qquad \text{F-15}$$

where,

$$\Gamma = (\nu/\omega)^{1/2} / (R_o \sin\beta) \qquad \text{F-16}$$

with the boundary conditions,

$$U = V = W = = 0 \qquad 1 \leq \eta, \; \sigma = 0 \qquad \text{F-17(a)}$$

$$\frac{\partial U}{\partial \sigma} = \frac{\partial W}{\partial \sigma} = 0 \qquad 1 \leq \eta, \; \sigma = \delta^+ \qquad \text{F-17(b)}$$

$$P = 0 \qquad 1 \leq \eta, \; \sigma = \delta^+ \qquad \text{F-17(c)}$$

$$\int_0^{\delta^+} U d\sigma = Q_o^+ \qquad 1 \leq \eta \qquad \text{F-17(d)}$$

F-17 (a) is the no slip condition, (b) is the condition for no momentum flux across the free surface, (c) states that the pressure at the free surface equals the pressure in the gas phase and (d) is the conservation of mass.

The solution to the above equation set has been found by Hinze and Milborn [13],

$$V_r = \left(\frac{\omega^2 Q^2 \sin\beta}{12 \pi^2 \nu r}\right)^{1/3} \qquad \text{F-18}$$

giving the average radial velocity and,

$$\delta = \left(\frac{3 \nu Q}{2 \pi \omega^2 r^2 \sin\beta}\right)^{1/3} \qquad \text{F-19}$$

giving the film thickness as a function of the radial distance. Using the present notation, and coordinate system, these equations are:

$$U = \left(1 - \frac{\cot\beta}{Fr}\right)\left(\delta^+ \sigma - (1/2)\sigma^2\right) \qquad \text{F-20(a)}$$

and

$$\delta^+ = \left(\frac{3 Q_o^+}{1 - \frac{\cot\beta}{Fr}}\right)^{1/3} \qquad \text{F-20(a)}$$

while the tangential and meridional velocities are given by,

$$V = \left(3\sin\beta - 2\frac{\cos\beta}{Fr}\right)\left((1/2)\delta^+ \sigma^2\right) \qquad \text{F-21}$$

$$- (1/6)\sigma^3$$

$$W = -(1/3)\left(1 - \frac{\cot\beta}{Fr}\right)\left(\delta^+ \sigma^3 - (1/4)\sigma^4\right) \qquad \text{F-22}$$

$$+ 2\delta^+ 3\sigma)$$

and the pressure by,

$$P = (\cot\beta + (1/Fr))(\delta^+ - \sigma) \qquad \text{F-23}$$

Figure 4, 5, and 6 illustrate the film thickness along the surface of the cone, and Figures 7, 8 and 9 show the average radial velocity along the cone surface.

CONCLUSION

Film and velocity profiles are computed from equations F-20 thru F-23. These are required in part II to evaluate the stability.

Acknowledgement

We would like to thank Mr. James A. Keane for his assistance in preparing the manuscript. This work was partially supported by LANL Contract #9-XQ3-4484F.

REFERENCES

[1] Long,R.; Roubicek,R.; Creed, J.; Holbrook,S.:Centrifugal Film Reactor-A New Concept for Cell Cultivation. Bioprocess Engineering 3 (1988), 73-77.

[2] Greespan, H.P.:The Theory of Rotatating Fluids. Cambridge University Press, London. 1968. 5-7, 271-288.

[3] Theodorsen,T.; Reiger,A.: Experiments on rotating plates, rods and cylinders at high speeds. N.A.C.A. T.N. (1947), 793.

[4] Read,P.L.: Dyanamics and Instabilities of Ekman and Stewartson Layers. Rotating Fluids in Geophysical and Industrial Applications. CISM, Italy,(1992), 49.

[5] Hopfinger,E.J.: General Concepts and Examples of Rotating Fluids. Rotating Fluids in Geophysical and Industrial Applications. CISM, Italy,(1992), 3.

[6] Mory,M.: Inertial Waves. Rotating Fluids in Geophysical and Industrial Applications. CISM, Italy, (1992), 175.

[7] Skelland, A.H.P.: Diffusional Mass Transfer. John Wiley & Sons, Inc., New York,1974.

[8] Emmert, R. E.; Pigford, R. L.:Chemical Engineering Progress 50 (1954) 87.

[9] Portalski, S.; Clegg, A. J: Chemical Engineering Science, 26 (1971) 773.

[10] Banerjee, S.; Rhodes, E.; Scott, D. S.: Chemical Engineeing Science, 22 (1967) 43.

[11] Bruin, S.: Chemical Egineering Science, 22 (1970) 1475-1485.

[12] Stainthorp, F.P.; Allen, J. M.: Trans. Inst. Chem. Engnrs. 43 (1958) T85.

[13] Hinze, J. O.; Milborn, H.: Journal of Applied Mechanics, (June/1950), 145-153.

[14] Lin, C. C.: The Theory of Hydrodynamic Stability. Cambridge University Press, London, 1955.

[15] Gregory, N.; Stuart, J. T.; Walker, W. S.: Phil. Trans. Roy. Soc. A. 248. (1955). 155-199.

[16] Lilly, D. K.: Journal of Atmospheric Sciences, 23. (1966),481-494.

[17] Caldwell, D. R.; Van Atta, C. W.: Journal of Fluid Mechanics, 44 (1970) 79-95.

[18] Bird, R. B.; Stewart, W. E.; Lightfoot, E. N.: Transport Phenomena. John Wiley & Sons, New York, 1960.

[19] Fulford, G. D.: Adv. Chem. Engng. 5 (1964) 151.

[20] Ruckenstein, E.; Berbente, C.: Int. J. Heat Mass Transfer, 11, (1968), 743-751.

[21] Javdani, K.; Chemical Engineering Science, 29, (1974), 61

[22] Pacheco, J. F.: Mass Transfer In A Thin Film Bioreactor: A Comparative Study On A Rotor Consisting Of 29° And 42° Half-Angle Conical Surfaces. Master's Thesis, New Mexico State University, May 1992.

[23] Rahman, M. M.; Faghri, A.; General Papers in Heat Transfer and Heat Transfer in Hazardous Waste Processing, HTD-Vol.212, ASME. 1992.

[24] Tatro, P. R.; Mollö-Christensen, E.L: Journal of Fluid Mechanics, 28, (1967), 531-544.

[25] Faller, A. J.; Kaylor, R. E.: Journal of Atmospheric Science, 23, (1966), 466-480.

[26] Long, R: Personal communication, 1994.

Industrial Mixing Fundamentals with Applications

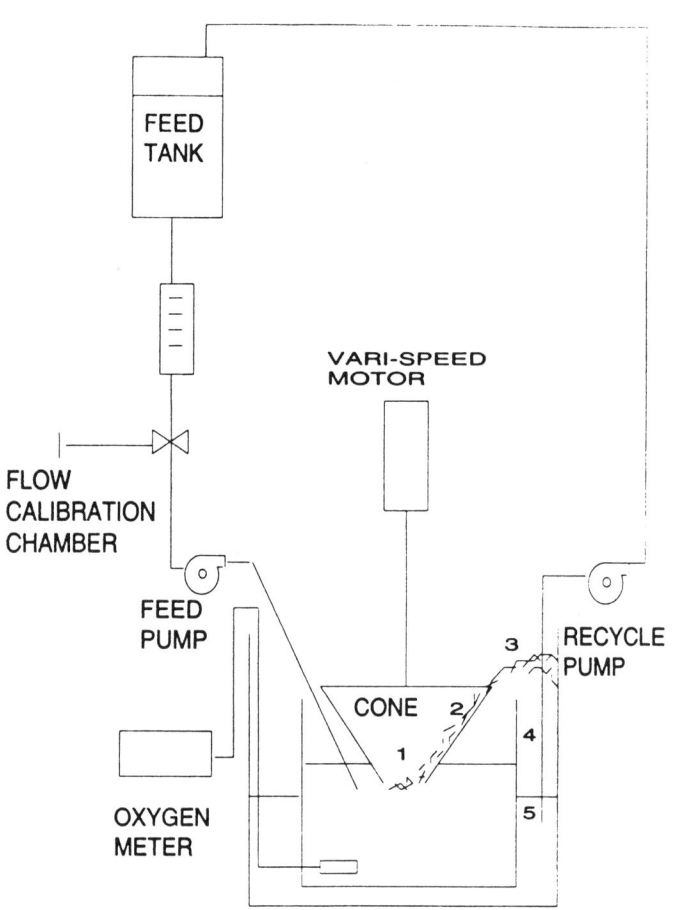

Figure 1: Flow apparatus and Region of Mass Transfer

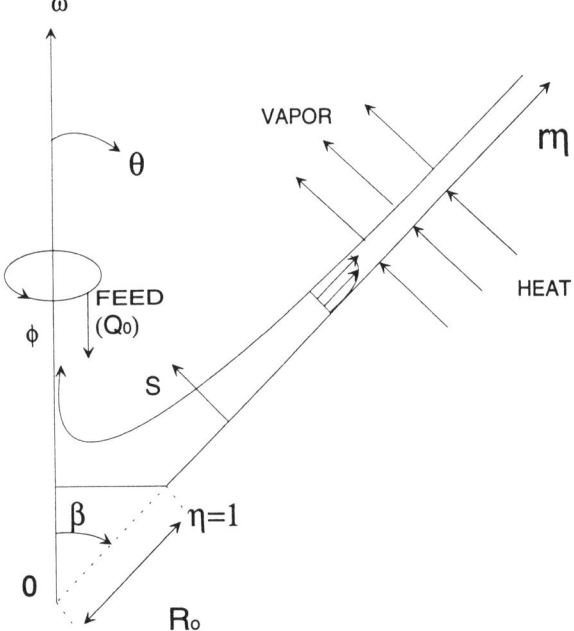

Figure 3 [11]

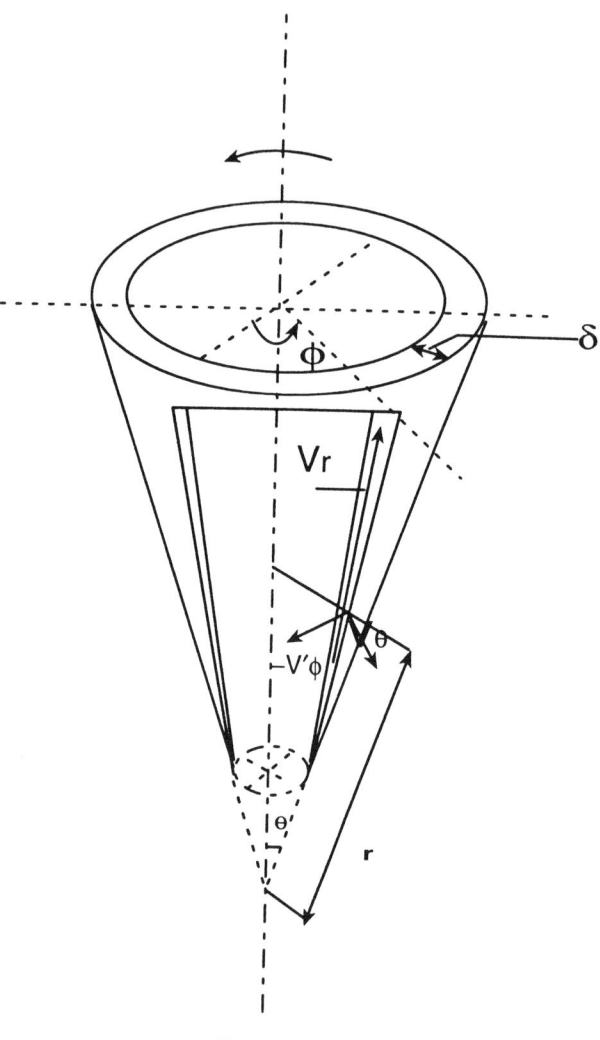

Figure 2 [11]

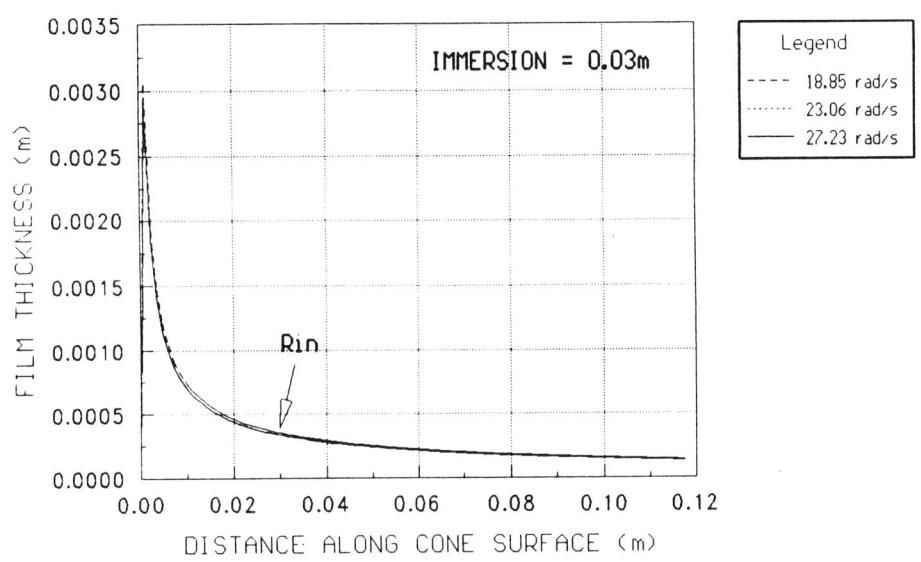

FIG 4: FILM THICKNESS ALONG CONE SURFACE

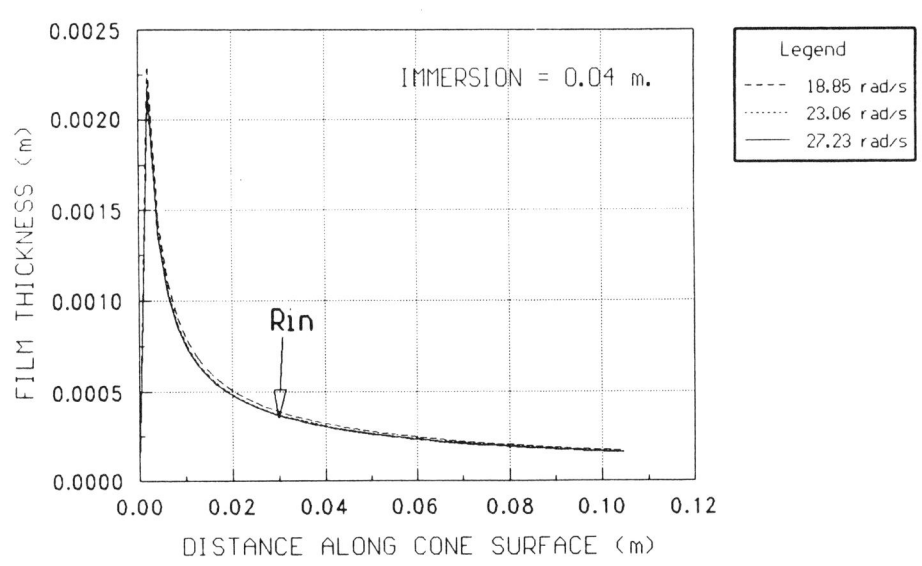

FIG 5: FILM THICKNESS ALONG CONE SURFACE

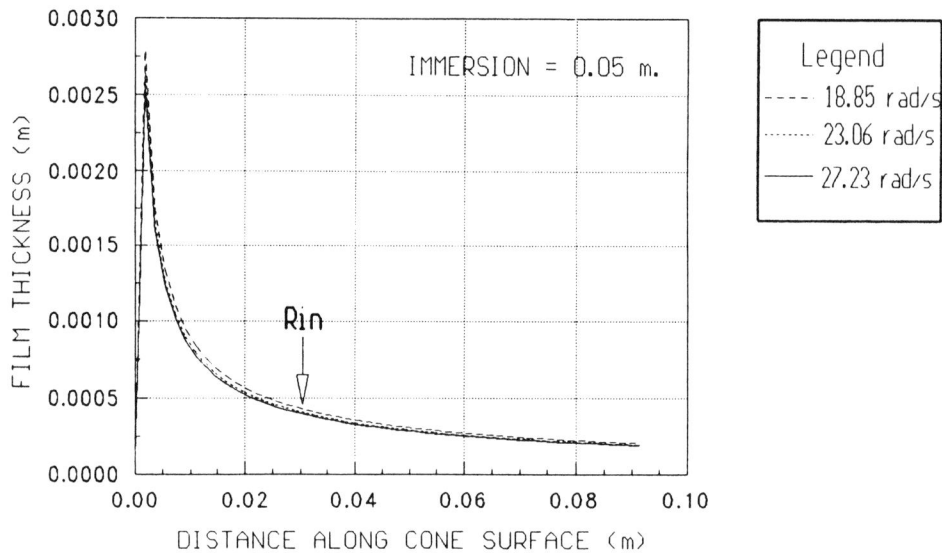

FIG 6: FILM THICKNESS ALONG CONE SURFACE

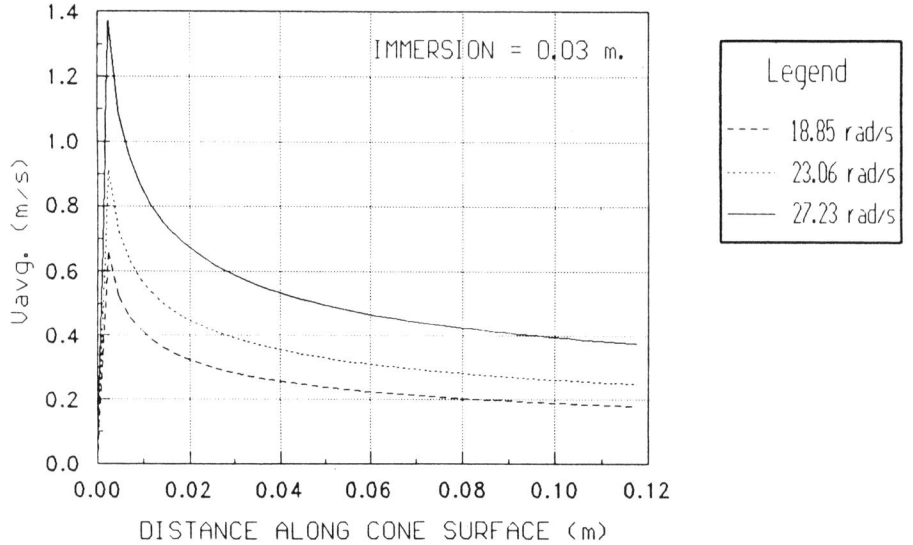

FIG 7: AVERAGE VELOCITY ALONG CONE SURFACE

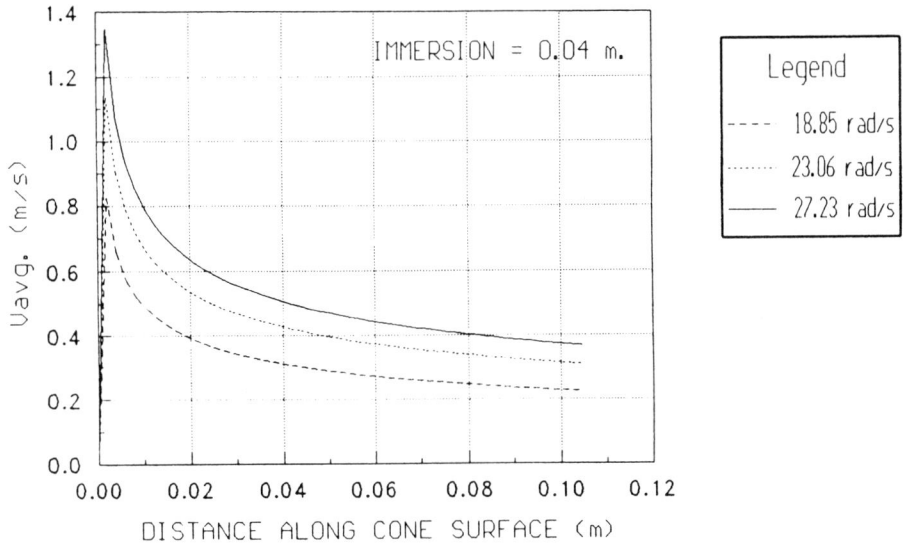

FIG 8: AVERAGE VELOCITY ALONG CONE SURFACE

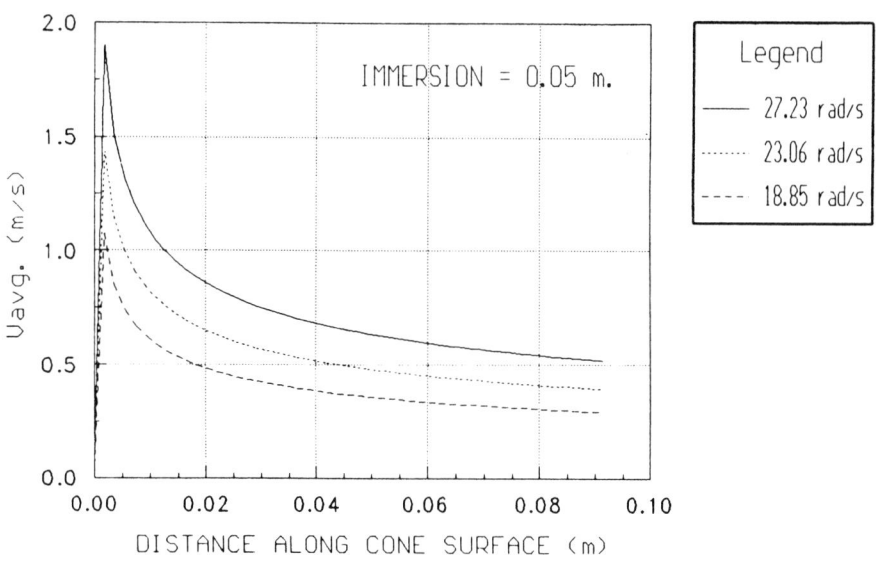

FIG 9: AVERAGE VELOCITY ALONG CONE SURFACE

Mass Transfer in a Laminar Rippling Film in a Conical Centrifugal Film Reactor

Part II: Stability Analysis for the Film Flow

Richard Long and Tushar Pattni
New Mexico State University, Department of Chemical Engineering, Las Cruces, NM 88003

In this section the film flow is evaluated by means of a linear stability analysis. This permits identification of the wave type of the ripples observed in the film. Evaluation of the Reynolds number, the Rossby number, and the Ekman number show that the waves are class A unstable primarily.

STABILITY OF THE EKMAN LAYER

The purpose of the Ekman layer analysis was to determine the wave characteristics required for the mass transfer analysis.

Experimental Background

Since instabilities depend only on the external flow field, they are a general feature of all rotational boundary layers. This fact was confirmed by the investigations of Theodorsen and Reiger [3], Smith [4], and Gregory, Stuart and Walker [5]. Due to this fact the experimental evidence cited is for motion in a cylindrical annulus and it is from the experiments of Tatro and Mollö-Christesen [24], as their measurements were quite accurate [2].([3,4,5,6] cited in Greenspan [2])

The basic experimental procedure of Tatro and Mollö-Chritensen [24] was to scan the flow with a hot wire anemometer. The results are shown in Figure 11. They show that there are two different types of instabilities of the boundary layer, designated as Class A and Class B instabilities according to the order in which they appear. The spacing of the waves is related to the depth of the boundary layer.

The stability problem is characterized in terms of Rossby and Reynolds numbers based on local measured values. (The Rossby number is a ratio of the convective acceleration to the Coriolis force.)

S-1

$$\epsilon = \frac{V1}{\Omega * r}$$

S-2

$$Re = \frac{\delta V1}{\nu}$$

$V1$ = relative azimuthal component of interior velocity
δ = local value of boundary layer thickness ($=4z/\pi$)
z = vertical coordinate z, of maximum radial velocity

The class A waves observed by Faller and Kaylor [4] are shown in Figure 11. The wavelength of these waves varied between $25(\delta)$ and $33(\delta)$. The wavelength of the class B waves is around $11.8(\delta)$ (Figure 12). Waves of the class A family develop first, and are sensitive to the value of ϵ. As ϵ increases, that is, as the convective acceleration increases, the disturbance
stops being confined only to the boundary layer and the effects are propagated to the interior.

The critical Reynolds number is more or less independent of the Rossby number, and is given by

Class A waves:
$$Re_A = 56.3 + 58.4\epsilon_A$$
Class B waves:
$$Re_B = 124.5 + 3.66\epsilon_B$$

According to the critical Re_A, the first boundary layer instability, that is, the formation of class A waves, may be expected at $Re=56$. Similarly class B waves may be expected at $Re=124$.

Theory of the Ekman Layer

The theoretical arguments proceed according to Greenspan [2]. The equations of motion are (in a coordinate system rotating at angular velocity Ω):

Dept. of Chemcial Engineering, New Mexico State University, Las Cruces, New Mexico, Dr. Richard Long & Tushar Pattni

F-1
$$\vec{\nabla}\cdot\vec{v}=0$$

$$\frac{\partial \vec{v}}{\partial t}+\vec{v}\cdot\vec{\nabla}\vec{v}+2\vec{\Omega}\times\vec{v}=-\frac{p}{\rho}\vec{\nabla}+g\vec{k}\rho+\nu\nabla^2\vec{v}$$

Now, assuming that the body force is conservative, i.e $F_b = P_f$, where
F_b = Body force
P_f = Force potential

the term containing the gravity term may be combined with the pressure term to give a new reduced pressure p' which is defined by the following expression:

$$p'=P+\rho P_f-\frac{1}{2}(\vec{\Omega}\times\vec{r})\cdot(\vec{\Omega}\times\vec{r})$$

Furthermore, making the equations of motion dimensionless using the following:

$$(\frac{\nu}{\Omega})^{1/2},\qquad (\frac{\nu}{\Omega})^{1/2}(\epsilon\Omega L)^{-1},\qquad \epsilon\Omega L$$

The fundamental equations become:

T-2
$$\vec{\nabla}\cdot\vec{v}'=0$$

$$Re(\frac{\partial \vec{v}'}{\partial t}+\vec{v}'\cdot\vec{\nabla}\vec{v}')+2\vec{k}\times\vec{v}'=-\vec{\nabla}p'+\nabla^2\vec{v}'$$

Let

T-3
$$\vec{v}'=\vec{v}'_I(\vec{r})+\mathcal{C}\vec{v}''(\vec{r},t)$$

$$p'=p'_1(\vec{r})+\mathcal{C}p''(\vec{r},t)$$

Where v_I' (r) is a steady laminar motion that satisfies the dimensionless equation of motion and all prescribed boundary conditions and $\mathcal{C}$ is a disturbance superposed on the mean flow. Substitution of the above expressions into (Equation(T-2) and linearization with respect to $\mathcal{C}$ results in

T-4
$$\vec{\nabla}\cdot\vec{v}''=0$$

$$Re(\frac{\partial \vec{v}''}{\partial t}+v_I\cdot\vec{\nabla}\vec{v}''+\vec{v}''\cdot\vec{\nabla}v_I)+2\vec{k}\times\vec{v}''=-\vec{\nabla}p''$$

$$+\nabla^2\vec{v}''$$

As stated by Greenspan [2], the extent of the fluid domain may be supposed to be infinite, bounded by a single plane wall.

The observed disturbances [17] are 2D vortex rolls and so we will consider wave-like perturbations relative to a coordinate system referring to the interior velocity (Figure 13). The x-coordinate measures along the bands and the y-coordinate measures along the normal. Therefore the interior velocity (dimensionless) is given by:

T-5
$$\hat{\theta}=\cos\alpha\hat{i}-\sin\alpha\hat{i}$$

where α is the positive angle of rotation. For the cylindrical coordinates, the x coordinate is directed circumferentially for a small α and y is a radial coordinate. The boundary layer velocity profile, therefore is:

T-6
$$v_1=v_{x1}\vec{i}+v_{y1}\vec{j}$$

$$v_1=(\cos\alpha-\exp(-z)\cos(\alpha+z))\vec{i}$$

$$-(\sin\alpha-\exp(-z)\sin(\alpha+z))\vec{j}$$

Introducing a stream function defined by the following since the perturbations in the form of 2D waves are independent of the x-axis:

T-7
$$v''_y=\frac{\partial \Psi''}{\partial z}\qquad v''_z=-\frac{\partial \Psi''}{\partial y}$$

where,

$$\Psi''(y,z,t) = \Psi(z) 0 \exp(ik(y-ct)) \qquad \text{T-8(a)}$$

$$v_x''(y,z,t) = V_x(z) \exp(ik(y-ct)) \qquad \text{T-8(b)}$$

is the assumed form of disturbance, and k = wave number.

By substituting the above expressions in the x-components of the momentum (Equation(T-2) and vorticity (Equation(T-10)),

$$Re\left(\frac{\partial \vec{v}'}{\partial t} + \vec{v}' \cdot \vec{\nabla} \vec{v}'\right) + 2\vec{k} \times \vec{v}' = -\vec{\nabla}p' + \nabla^2 \vec{v}' \qquad \text{T-2}$$

and

$$\frac{\partial B}{\partial t} + \vec{\nabla} \times [(\epsilon B + 2\vec{k}) \times \vec{v}'] = -E\vec{\nabla} \times \vec{\nabla} \times B \qquad \text{T-10}$$

where

$$B = \vec{\nabla} \times \vec{v}'$$

one obtains the following O.D.E.'s [2]:

$$\frac{d^2 V_x}{dz^2} - k^2 V_x + 2\frac{d\Psi}{dz} - ik\,Re[(v'_{y1} - c)V_x \qquad \text{T-11}$$

$$-\Psi \frac{dv'_{x1}}{dz}] = 0$$

$$\left(\frac{d^2}{dz^2} - k^2\right)^2 \Psi - ik\,Re\left[(v'_{y1} - c)\left(\frac{d^2}{dz^2} - k^2\right)\Psi - \Psi \frac{d^2 v'_{y1}}{dz^2}\right] - 2\frac{dV_x}{dz} = 0$$

with the following boundary conditions,

$$V_x = 0, \quad \Psi = \frac{d\Psi}{dz} = 0 \quad at\ z=0\ (the\ plate) \qquad \text{T-13}$$

$$(V_x, \Psi, \frac{d\Psi}{dz}) \to 0 \qquad as\ z \to \infty \qquad \text{T-14}$$

(Equations(T-11 and T-12) with the boundary conditions (Equations(T-13 & T-14) is an eigenvalue problem in which c, the phase speed, is the characteristic value Ψ, and U are the eigen functions and k, Re and α are parameters. Although this eigenvalue problem is for a flat plate, the analysis is applicable to the configuration in the present study (the cone) due to the fact that the film formed is very thin and the Ekman characteristics are a result of the flow field only. Now, c is a single valued complex function of the basic parameters, for each mode. Letting the imaginary part of c be Im c(α,k,Re), then,

Im c(α,k,Re) = 0 is the stability surface.

When Im c(α,k,Re) > 0, there is a positive amplification and the wave is unstable. Therefore, Im c(α,k,Re) = 0 is the boundary between stable and unstable conditions. The objective then becomes to determine the smallest values of Re and the associated values of α and k on this curve since this is the Re at which the instability is first encountered. Laboratory experience suggests that the instabilities set in when viscous effects are relatively small compared to inertial effects i.e Re $\to$ infinity (Figure 14). Therefore the inviscid form of (Equations(T-11 and T-12) is

$$(v'_{y1} - c)\left(\frac{d^2}{dz^2} - k\right)\Psi - \Psi \frac{d^2 v'_{y1}}{dz^2} = 0 \qquad \text{T-15}$$

$$V_x = \frac{\Psi}{v'_{y1} - c} \frac{dv'_{x1}}{dz} \qquad \text{T-16}$$

This is the classical inviscid stability problem and the details can be found in C.C. Lin's book [14]. In this approach the rotational effects are completely suppressed and the (Equations(T-15 & T-16) uncouple as the limit is approached i.e Re=infinity. Detailed investigations (Stuart [15]) of the inviscid instabilities indicate their consistency with the observations, except that the wave number predicted is too large. Thus, viscosity has a minor influence in the determination of α (since the observed and theoretical orientation angles are relatively close), but it seems that it is essential to the selection of k. The full viscous form (Equation(T-12) without the 2dU/dz term is identical to the Orr-Sommerfeld equation. Numerical work (Lilly [16]) confirms that the Class B waves are governed mainly by this O-S equation. Inviscid analysis does not account for the Class A waves since these waves are from an overturning that involves the Coriolis and shear forces. Again, Lilly [16] exhibited unstable waves associated with the Coriolis force. The waves disappear at high Re, when the Coriolis force and the shear forces uncouple. Figure 13 show the results of Lilly's computations. The diagrams are of the planes of surfaces of constant Im c(α,k,Re) at various Re. They show the regions of stability, instability (shaded) and the point of maximum growth rate. The diagrams at higher Re (110,150,500) show a shift of the unstable zone to angles of positive orientation and to small phase speeds. Both these signs are symptomatic of the emergence of Class B modes. Figure 14 shows the growth rates as a function of Re. From the diagram, Class A modes appear at a critical Re=55 and Class B at a critical Re=110. This is in substantial agreement with experimental evidence.

As stated earlier, the Ekman number is a measure of how the typical viscous force compares to the Coriolis force. To calculate the Ekman number for the flow in the present case, it will be defined thus:

T-17
$$E = \frac{\nu}{\omega * r^2}$$

where w is the angular velocity at radius r. This angular velocity w can be approximated by Ω. Since the Ekman number is a function of the radial distance, it is necessary to use values of r that are in the domain that is being considered. The immersion depth would not be a prudent location to use for the calculation of the entrance Ekman number, because at this location the flow is not defined by the preceding equation. The locations chosen, therefore, were at the mid-point and at the edge of the conical surface (along the radial direction), above the depth of immersion. In addition to these two locations, a point, Rin (see Figures 4, 5 and 6), was selected. This point represented where the thin film domain began. From the graphs of the film thickness, Rin was selected at 0.03 m above the immersion depth along the cone surface. The Ekman numbers calculated for these locations are shown in Table ES-1. The Rossby and Reynolds numbers are defined as follows (with v_r being the average radial velocity [13]):

T-18
$$\epsilon = \frac{v_r}{r * \omega}$$

T-19
$$Re = \frac{v_r * \delta}{\nu}$$

The values of the Rossby and Reynolds numbers for the above mentioned locations are indicated in Table ES-1.

From the range of the Reynolds number it is evident that both Class A and Class B waves should be present (see Table EX-2 and Figure 15).

CONCLUSION

Figure 15 shows clearly that we have both class A and Class B waves present in the range of operations evaluated in this set of data. This information is used in part III to evaluate the enhancement of mass transfer due to the ripples in the film

Acknowledgment

We would like to thank Mr. James A. Keane for his assistance in preparing the manuscript. This work was partially supported by LANL Contract #9-XQ3-4484F.

References

See References, part I.

SUMMARY OF EKMAN LAYER INSTABILITIES [17]

Type of Instability	Quantity	THEORY		EXPERIMENTAL		
		Faller	Lilly	Faller	Tatro	Caldwell
Class A	Critical Reynold Number	55	55	<70	56.3 + 116.8* Ro	56.7
	Wavelength (Ekman Depths)	24	21	22-33	27.8 + 2.0	
	Inclination of rolls	-15°	-20°	5° to -20°	0 - -8°	
	Geostrophic Velocity	0.5	0.57		0.16	
Class B	Critical Reynolds Number	118	110	125 + 5	124.5 + 7.32* Ro	
	Wavelength (Ekman Depths)	11	11.9	10.9	11.8	
	Inclination of rolls	10°--12°	8°	14.5 + 2°	14.8° + 0.8°	
	Geostrophic Velocity	0.33	.094	0.023	0.034	

TABLE ES-1 (f): IMMERSION=4cm., ANGULAR VELOCITY=27.23 rad/s

Distance along Cone Surface (m)	0.03 (Rin)	0.055	0.10
Ekman Number	2.64E-5	7.84E-6	2.37E-6
Rossby Number	0.42	0.22	0.10
Reynolds Number	215	88	61

TABLE ES-1 (g): IMMERSION=5cm., ANGULAR VELOCITY=18.85 rad/s

Distance along Cone Surface (m)	0.03 (Rin)	0.05	0.09
Ekman Number	3.81E-5	1.37E-5	4.22E-6
Rossby Number	0.48	0.27	0.11
Reynolds Number	184	120	58

TABLE ES-1 (h): IMMERSION=5cm., ANGULAR VELOCITY=23.06 rad/s

Distance along Cone Surface (m)	0.03 (Rin)	0.05	0.09
Ekman Number	3.11E-5	1.11E-5	3.46E-6
Rossby number	0.53	0.27	0.12
Reynolds Number	233	139	74

TABLE ES-1 (i): IMMERSION=5cm., ANGULAR VELOCITY=27.23 rad/s

Distance along Cone Surface (m)	0.03 (Rin)	0.05	0.09
Ekman Number	2.64E-5	9.45E-6	2.92E-6
Rossby Number	0.59	0.29	0.14
Reynolds Number	299	179	98

TABLE ES-1 (a): IMMERSION = 3cm., ANGULAR VELOCITY = 18.85 rad/s

Distance along Cone Surface (m)	0.03 (Rin)	0.06	0.11
Ekman Number	3.81E-5	9.51E-6	2.83E-6
Rossby Number	0.31	0.12	0.05
Reynolds Number	116	42	17

TABLE ES-1 (b): IMMERSION = 3cm., ANGULAR VELOCITY = 23.06 rad/s

Distance along Cone Surface (m)	0.03 (Rin)	0.06	0.11
Ekman Number	3.11E-5	7.78E-6	2.31E-6
Rossby Number	0.34	0.14	0.06
Reynolds Number	147	60	24

TABLE ES-1 (c): IMMERSION = 3cm., ANGULAR VELOCITY = 27.23 rad/s

Distance along Cone Surface (m)	0.03 (Rin)	0.06	0.11
Ekman Number	2.64E-5	6.56E-6	1.97E-6
Rossby Number	0.45	0.18	0.08
Reynolds Number	227	90	37

TABLE ES-1 (d): IMMERSION = 4cm., ANGULAR VELOCITY = 18.85 rad/s

Distance along Cone Surface (m)	0.03 (Rin)	0.055	0.10
Ekman Number	3.81E-5	1.13E-5	3.43E-6
Rossby Number	0.38	0.17	0.08
Reynolds Number	131	54	37

TABLE ES-1 (e): IMMERSION = 4cm., ANGULAR VELOCITY = 23.06 rad/s

Distance along Cone Surface (m)	0.03 (Rin)	0.055	0.10
Ekman Number	3.11E-5	9.26E-6	2.80E-6
Rossby Number	0.42	0.19	0.09
Reynolds Number	179	74	52

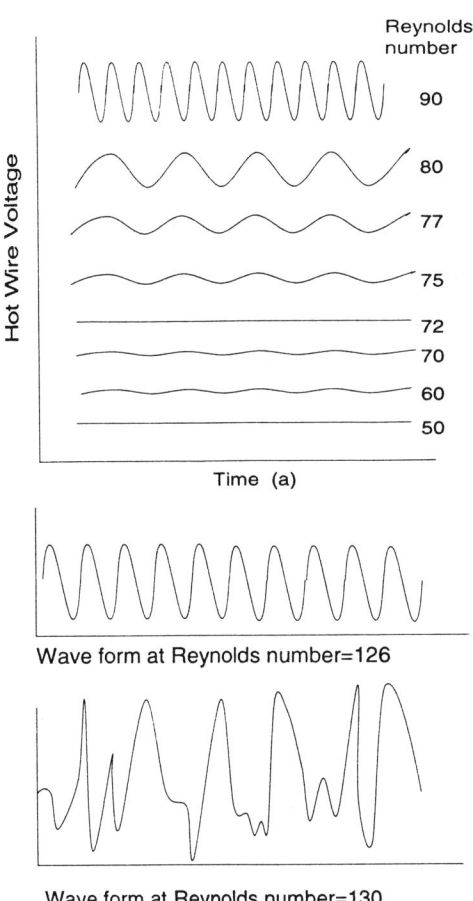

Figure 10: (a) Record of Hot Wire Voltage by Tatro and Mollo-Christensen showing onset of Instability and Class A waves. (b) Class B waves [24](cited in [2]) resketched here for illustration.

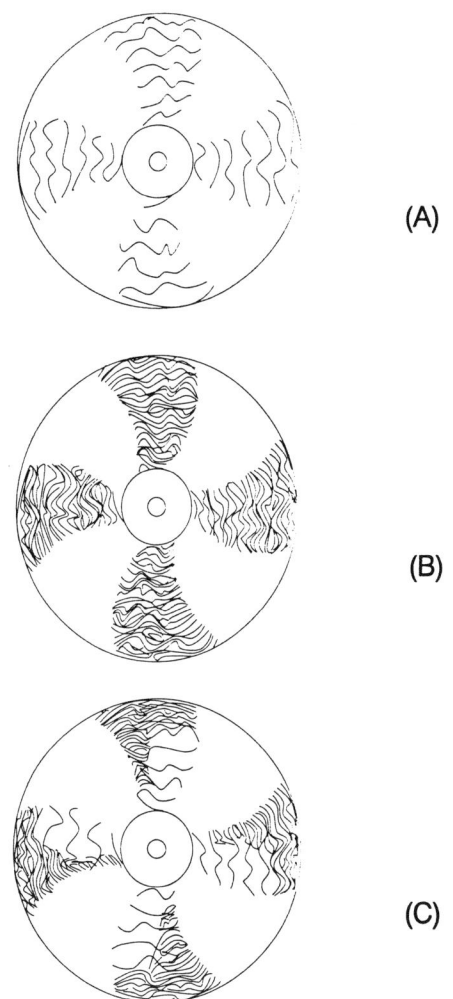

Figure 11 Ripples sketched from photos by Faller and Kaylor [25] (cited in [2]) (A) Class A waves (B) Class B Waves (C) Class A and Class B Waves

78 Industrial Mixing Fundamentals with Applications   AIChE SYMPOSIUM SERIES

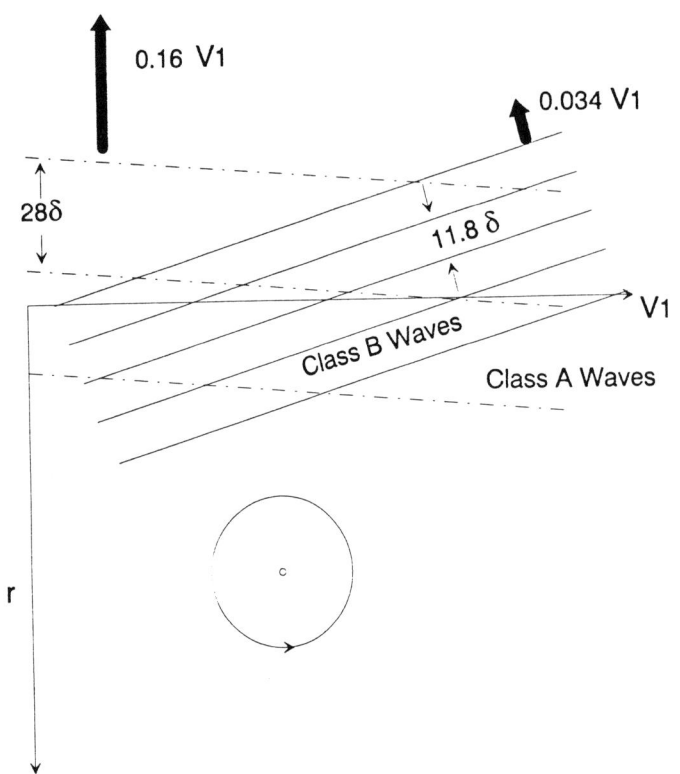

Figure 12: The Two Instablility Classes and their Wave Geometry (Sketched from[2]) for illustration)

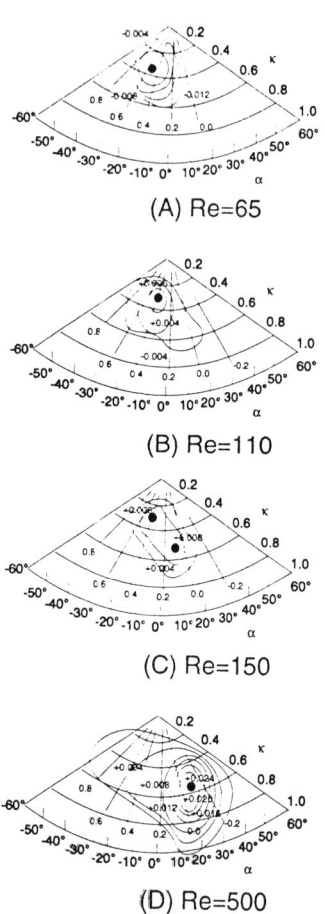

(A) Re=65

(B) Re=110

(C) Re=150

(D) Re=500

Figure 13 Growth Rates and Phase Velocities. Cited in [2], and sketched from [16] for illustration

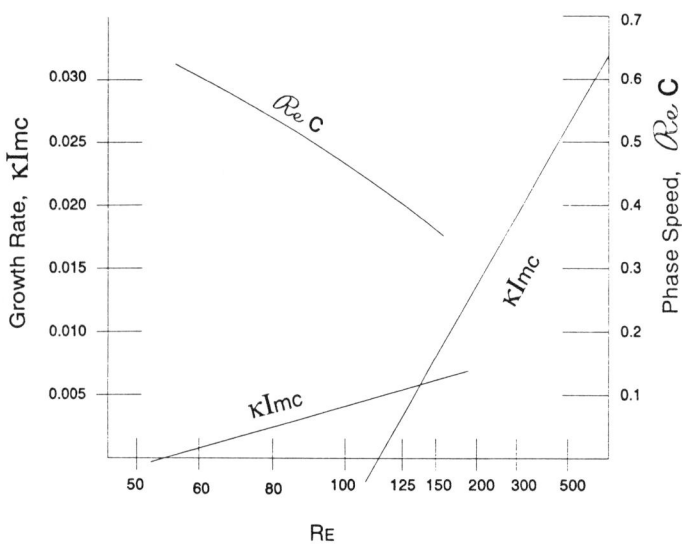

Figure 14 Growth Rate of Instability as a Function of Reynolds Number cited in[16] and sketched here for illustration.

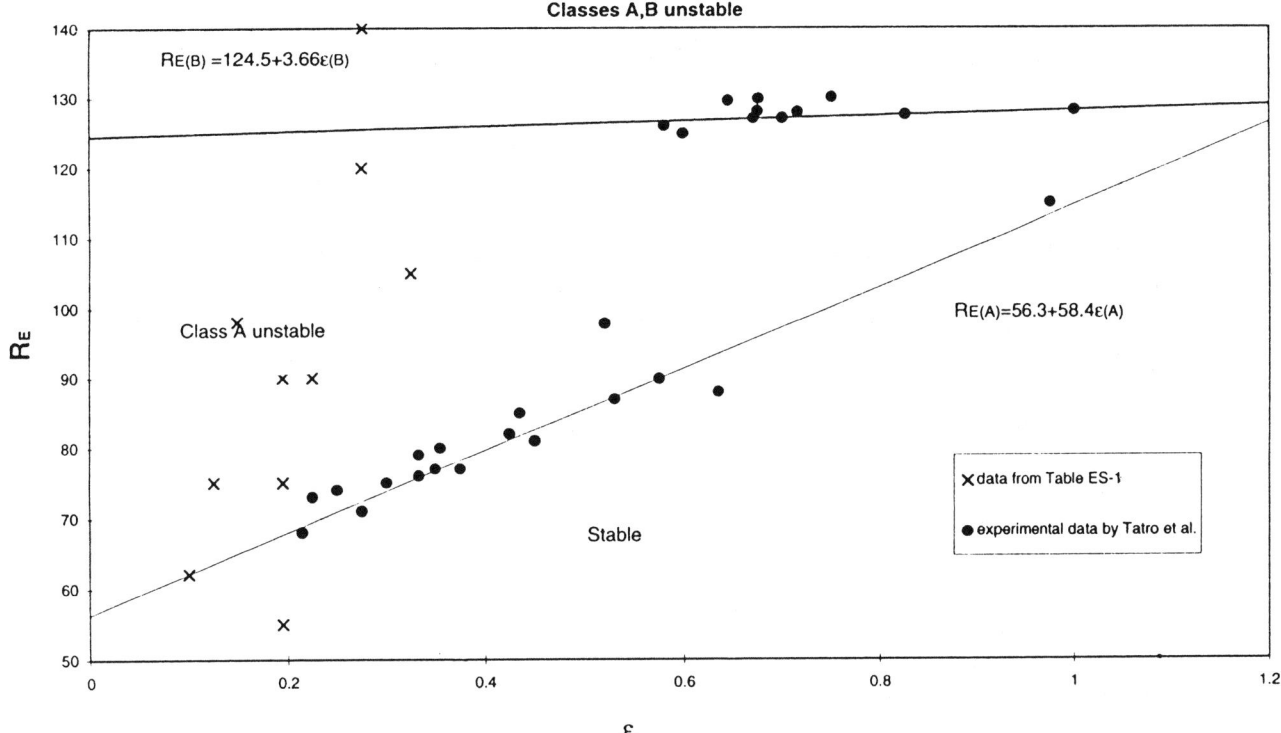

Figure 15: Critical Reynolds Number vs. Rossby Number for Class A and Class B instabilities,

lines sketched from[23] data from out Table ES-1 superimposed to show condition of our experiments.

Mass Transfer in a Laminar Rippling Film in a Conical Centrifugal Film Reactor

Part III: Enhanced Mass Transfer in Ripple Flow

Richard Long and Tushar Pattni
New Mexico State University, Department of Chemical Engineering, Las Cruces, NM 88803

The film flow analysis of Part I and the stability analysis of Part II are applied to compute the mass transfer coefficient for the film flow on the cone. The results show that there is an enhancement of mass transfer by about a factor of 2 for the rippling flow. It is also shown that the film mass transfer on the cone is about 15% - 30% of the overall mass transfer in this device as configured.

MASS TRANSFER ANALYSIS

Now, the region under consideration is the thin laminar film (TLF) formed on the cone surface beyond the initial entrance zone. From the section on the fluid flow, this (laminar) region began approximately 0.03 m. above the immersion depth (see Figures 6, 7 and 8). The concentration profiles formed in the TLF are developed in parallel to the temperature profiles in Bruin's [11] heat transfer analysis, except with a change in the boundary conditions.

A stream function Ψ^+ [18] is introduced into the Navier-Stokes equation with constant density and viscosity, and the continuity equation. This yields the following equations:

$$\frac{\partial^3(\Psi^+)}{\partial\sigma^3} = -\eta^2 \sin^2\beta \qquad \text{M-1}$$

and

$$\left(\frac{Sc}{\eta}\right)\left(\frac{\partial(\Psi^+)}{\partial\sigma}*\frac{\partial\Theta}{\partial\eta} - \frac{\partial(\Psi^+)}{\partial\eta}*\frac{\partial\Theta}{\partial\sigma}\right) \qquad \text{M-2}$$

$$= \sin\beta \frac{\partial^2\Theta}{\partial\sigma^2}$$

where,

$$\Theta = (C_o - C)/(C_o - C_b) \qquad \text{M-3}$$

$$\Psi^+ = \Psi/(\omega\nu)^{1/2}(R_o)^2 \qquad \text{M-4}$$

$$U = \frac{1}{(\eta\sin\beta)^2}\frac{\partial(\Psi^+)}{\partial\sigma} \qquad \text{M-5}$$

and,

$$V = -\frac{1}{\eta\sin\beta}\frac{\partial(\Psi^+)}{\partial\eta} \qquad \text{M-6}$$

Introducing a new function F (defined below), which is only a function of σ to the differential equations, they become:

$$F = \Psi^+/\eta^2$$

$$\sin\beta\left(\frac{\partial^2\Theta}{\partial\sigma^2}\right) + 2F\left(\frac{\partial\Theta}{\partial\sigma}\right)Sc = 0 \qquad \text{M-7}$$

and

$$\frac{\partial^3 F}{\partial\sigma^3} + \sin^2\beta = 0 \qquad \text{M-8}$$

with the following boundary conditions:

$$\sigma = 0, F = F_\sigma = 0 ; \Theta = 1$$
$$\sigma = \delta^+(\eta), F_{\sigma\sigma} = 0 ; \Theta = 0$$

The solution to equation M-8,9 is

M-10

$$\Theta(s,r) = \int_0^\alpha \exp(-Sc\delta^{+4}\sin\beta((\frac{z^3}{3})-(\frac{z^4}{12})))dz$$

$$\div (\int_1^0 \exp(-Sc\delta^{+4}\sin\beta(\frac{z^3}{3}-\frac{z^4}{12}))dz)$$

Where, $\alpha=\sigma/\delta^+$ and z is the dummy integration variable.

On doing a mass balance around the film surface, a distance r from the immersion one obtains the following relationship:

M-11

$$\sin\beta \frac{d}{dr}(\int_0^\delta \rho v_r ds) = D\frac{dc}{dr}\bigg|_{s=\delta} \sin\beta$$

Introducing a "mass number",

$$M = (\frac{\nu}{\omega})^{1/2}\frac{C_o-C_b}{\rho}$$

the dimensionless form of equation M-11 becomes,

M-12

$$\frac{d}{d\eta}(\int_0^{\delta^+} \eta U d\sigma) = -\frac{M}{Sc\sin\beta}\Theta$$

This equation is in the same form as Bruin's [11] (Equation(33)) for heat balance. The Sherwood number is defined as follows:

M-13

$$Sh_{loc} = \frac{k_{l(loc)}r}{D}$$

and the mass transfer coefficient is

M-14

$$k_{l(loc)} = \frac{n_{Ao}}{C_A'} = \frac{D(\partial C/\partial s)|_{s=0}}{C_A'}$$

where n_{Ao} = Mass flux of component A, and $C_A' = (C_o-C_b)$. Using equations M-13 and M-14, one obtains

M-15

$$Sh_{loc} = \frac{r(\partial C/\partial s)|_{s=0}}{C_o-C_b}$$

Therefore in order to calculate the Sherwood number the concentration gradient is required. This is obtained by differentiating equation M-10 resulting in:

M-16

$$(\Theta_\sigma)|_{\sigma=\delta^+} = \frac{\exp(-1/4Sc\delta^{+4}\sin\beta)}{\delta^+\int_0^1 \exp(-Sc\delta^{+4}\sin\beta(\frac{z^3}{3}-\frac{z^4}{12}))dz}$$

The expression M-16 was evaluated for various values of the dimensionless film thickness, and Sc=575.45 (Table M1).

The data was fitted with a linear profile for the temperature gradient ($1/\delta^+$) with a good fit (Figure 16). The equation of the line was $-0.81 + 1.156*(1/\delta^+)$, with the Goodness of Fit coefficients as follows:

C.O.D : 0.9861
Corrl : 0.9930
M.S.C : 3.8759

Therefore the concentration gradient can be replaced by $1/\delta^+$ when solving the dimensionless film surface mass balance equation to obtain the relation between the dimensionless film thickness, δ^+, and the dimensionless radial coordinate, η (Bruin [11]).
This relationship is

$$\delta^+ = \delta_F^+ \eta^{-2/3}$$

Using this profile, one obtains the following expression for the dimensionless local Sherwood number;

M-17

$$Sh_{loc} = R_o(\frac{\omega}{\nu})^{1/2}\eta(\Theta_\sigma)|_{\sigma=\delta^+}$$

$$Sh_{loc} = R_o(\frac{\omega}{\nu})^{1/2}\frac{\eta}{\delta^+}$$

Substituting into the local Sherwood number results in the following expression:

M-19

$$Sh_{loc} = R_o(\frac{\omega}{\nu})^{1/2}\frac{\eta^{5/3}}{\delta_F^+}$$

Therefore the local mass transfer coefficient becomes

M-20

$$k_{l(loc)lin} = \frac{Sh_{loc}D}{\eta R_o} = \frac{D}{\delta_F^+}(\frac{\omega}{\nu})^{1/2}\eta^{2/3}$$

The average Sherwood number can be found by integrating over the length of the non-immersed cone.

$$Sh_{av} = \frac{\int_{\eta_i}^{\eta_{out}} R_o (\frac{\omega}{\nu})^{1/2} \eta^{5/3} d\eta}{\int_{\eta_i}^{\eta_{out}} \eta \, d\eta}$$

M-21

resulting in,

$$Sh_{av} = \frac{3}{8} \frac{R_o}{\delta_F^+} (\frac{\omega}{\nu})^{1/2} \frac{\eta_{out}^{8/3} - \eta_i^{8/3}}{\eta_{out} - \eta_i}$$

M-22

Similarly, the average mass transfer coefficient can be calculated from the following expression

$$k_{l(av)} = \frac{\int_{\eta_i}^{\eta_{out}} \frac{D}{\delta_F^+} (\frac{\omega}{\nu})^{1/2} \eta^{2/3} d\eta}{\int_{\eta_i}^{\eta_{out}} \eta \, d\eta}$$

M-23

which results in,

$$k_{l(av)} = \frac{3}{5} \frac{D}{\delta_F^+} (\frac{\omega}{\nu})^{1/2} \frac{\eta_{out}^{5/3} - \eta_i^{5/3}}{(\eta_{out} - \eta_i)}$$

M-24

The mass transfer coefficients calculated using the above expression and a value for the diffusivity of oxygen into water $D = 7.99E\text{-}6 \, m^2/hr$, at 20° are in units of m/hr. To convert them to the more conventional units of hr^{-1}, they are multiplied by the contact length of the cone L_c ($L_c = A'/V'$, where A' = effective exposed surface area of cone and V' = effective exposed volume of cone). These mass transfer coefficients are given in Table M2.

Mass Transfer in Wavy Laminar Flow

At Reynolds numbers between 100 and 800 defined as follows, (using the flow rate per contact length, i.e L_c:

$$Re_l = \frac{L_c}{\nu}$$

where, Q = Volumetric flow rate

Banerjee et al [10], have proposed a model which is applicable to rising films with waves. The model is based on the fact that eddies are formed in the film due to the passage of waves and it is used to calculate the enhanced mass transfer coefficient associated with the wavy film formed on the inside of the CFP. Reynolds numbers using the above formula for the present case are given in Table M3.

The Eddy Model

A circulating eddy is associated with each wave (Figure 17). Then the average time period between each wave is given by the following relationship of the flow parameters:

$$t = \frac{\overline{\lambda}}{\overline{u_w} - \overline{u_s}}$$

where,
u_w = average velocity of wave
u_s = average velocity at the film surface
$\overline{\lambda}$ = average wave length

This eddy comes to a distance H from the surface of the film, depending on the liquid properties and the velocity of the wave, where it is renewed. The degree to which this renewal occurs is a function of H. In other words, the closer the eddy comes to the surface, the higher is the degree of renewal resulting in a higher mass transfer rate. To calculate the average distance $\overline{H}$, within which the eddies come to the surface of the film, Banerjee et al [10], have proposed the following relationship:

$$\overline{H} = (\frac{\phi \nu}{\rho u_w^3})^{1/2}$$

where, Φ = surface tension of the liquid.

Using the above mentioned parameters in the region of interest, that is for Reynolds numbers between 100 and 800, Banerjee et al's [10] relation for approximating the mass transfer coefficient is as follows:

$$k_{lav(w)} = \frac{0.7 \, D^{5/8}}{H^{1/4} t^{3/8}}$$

The theoretical mass transfer coefficients, $k_{lav(w)}$, again, are converted to the conventional units using the contact length. The values obtained for the present case are given in Table M4.

The value of the diffusivity used was $7.99E\text{-}6 \, m^2/hr$ as in the case for the mass transfer coefficient for the non-wavy flow. The velocity at the surface of the wavy film was calculated using the empirical relationship found by Fulford [19], which was

$$u_s \approx 1.5 \, u_{av}$$

u_{av} was determined by integrating the equation for the average velocity as a function of the radial coordinate, over the exposed radial distance.

$$u_{av} = 1.5 (\frac{\omega^2 Q^2}{12 \pi^2 \nu})^{1/3} \frac{R_{out}^{2/3} - R_i^{2/3}}{R_{out} - R_i}$$

u_w, the average wave velocity, was calculated using the relationship found by Stainthorp and Allen [12],

$$u_w = 2.95 \, Re^{1/2}$$

Due to lack of data, the Reynolds numbers from Table ES-1 were used as an indication to determine the average wavelength. It was assumed that most of the waves were of Class A and hence the average wavelength was approximated by 25δ.

The increase in the mass transfer coefficient due to the wavy flow is given in Table M5.

Comparison with Overall Mass Transfer Coefficient

The overall mass transfer coefficients, for a 5 cm. immersion, obtained for the CFR by Long et al [1], are given in Table M6.

Based on these values, a comparison to the overall mass transfer coefficient can be made. The percentage of the total mass transfer coefficient that is due to the TLF on the interior surface of the CFR is shown in Table M7 (for the 5 cm. immersion)

DISCUSSION

Although in solving the fluid flow equations various simplifying assumptions were made, they did not affect the final result, that is the increase in the mass transfer coefficient. The increase in the mass transfer coefficient for the wavy laminar flow was within the expected range of approximately 200 per cent.

From Figures 6, 7 and 8 it is evident that as the radial coordinate increases, the average film thickness not only decreases but its value for the different angular velocities is relatively close, especially at higher values of the radial coordinate. This is probably due to the fact that the values of the angular velocities were relatively similar. However, the radial velocity is much more sensitive to the angular velocity and there was a significant difference in the average radial velocity for the three different angular velocities (Figures 9, 10 and 11).

The Reynolds numbers (Re_v) based on the velocity are tabulated in Table ES-1. Based on Figure 15, near the entrance both Class A and Class B waves are present. As the radial coordinate increases, the Reynolds number decreases and the flow regime consists primarily of Class A waves. As the immersion depth increases there is a higher proportion of Class A and Class B waves as opposed to a Class A waves only. As the wavelength of the waves depends on the type of wavy flow, and since no data was available the assumption was made that the average wavelength was 25δ. This wavelength corresponds to small wavelength Class A waves. Class B waves are generally of order 11.8δ. The Reynolds number based on the flow per contact length, Re_l, are in the range where Banerjee et al's [10] empirical model to calculate the mass transfer coefficient would be applicable.

Critical to this and other models [20,21] for determining the mass transfer coefficient, is the wavelength and the celerity of the waves. Again, since no data was available these parameters had to be estimated as has been previously mentioned. A shorter wavelength would result in a shorter time period between eddies and hence a higher rate of surface renewal of the eddies. This would manifest itself in a higher mass transfer coefficient. Therefore the choice of the wavelength in this case, may not accurately portray the actual parameter. This is probably manifested the most for the case of the 4 and 5 cm. immersion depths, as in this configuration there is probably a higher percentage of Class B waves. That is probably why the mass transfer coefficient for the for the wavy film, $k_{la(w)}$, seems to be decreasing for higher angular velocities. The increase in the mass transfer coefficient due to the wavy flow was 140 - 250 %. This result is more or less consistent with experimental results [27].

It was found that approximately 15-30 % of the mass transfer in the CFR could be occurring in the region of the TLF on the interior surface of the cone. A similar thin film exists on the outer surface, although it is complete at higher angular velocities. Therefore the percentage of the mass transfer occurring on the conical surface may be as high as 50 %.

CONCLUSIONS

1. Since 15-30% of the mass transfer seems to be taking place in the region under study, i.e regime 2, 70-85% must be occurring in the rest of the reactor.

2. Increasing the angular velocity will result in a higher proportion of Class B waves with a shorter wavelength. This will result in a higher degree of surface renewal of the circulating eddy which may result in a higher mass transfer coefficient. The present model for the mass transfer coefficient, however, may not be applicable due to the higher Reynolds number. Additionally, it was also observed by Long [1], that after certain higher angular velocities, the integrity of the thin film is compromised and the film is not continuous.

3. Pertinent experimental data is required for more accurate analysis.

Acknowledgement

We would like to thank Mr. James A. Keane for his assistance in preparing the manuscript. This work was partially supported by LANL Contract #9-XQ3-4484F.

References
See References, Part I

TABLE M1: CONCENTRATION GRADIENTS AT THE SURFACE OF THE LAMINAR FILM. (Sc = 575.45 ß = 0.698)

$1/\delta^*$	Θ_σ
1.149	0.2509
1.176	0.2973
1.250	0.4329
1.333	0.7742
1.538	0.8734
1.818	1.645
2.222	2.0106
2.857	2.7526
4.000	3.9601
6.667	6.6600

TABLE M2: MASS TRANSFER COEFFICIENTS FOR THE DIFFUSION OF OXYGEN INTO THE LAMINAR FILM (20°C)

IMMERSION (m)	ROTATIONAL SPEED (rad/s)	$k_{l(av)}$ (hr^{-1})
0.03	18.85	0.64
0.03	23.06	0.64
0.03	27.23	0.67
0.04	18.85	0.64
0.04	23.06	0.72
0.04	27.23	0.72
0.05	18.85	0.52
0.05	23.06	0.52
0.05	27.23	0.57

TABLE M3: REYNOLDS NUMBERS USING THE FLOW RATE PER CONTACT LENGTH

IMMERSION (m)	ROTATIONAL SPEED (rad/s)	Re_l
0.03	18.85	320
0.03	23.06	414
0.03	27.23	574
0.04	18.85	425
0.04	23.06	558
0.04	27.23	730
0.05	18.85	572
0.05	23.06	728
0.05	27.23	923

TABLE M4: THEORETICAL MASS TRANSFER COEFFICIENTS FOR THE WAVY FILM

IMMERSION (m)	ROTATIONAL SPEED (rad/s)	$k_{lav(w)}$ (hr^{-1})
0.03	18.85	1.82
0.03	23.06	1.82
0.03	27.23	1.77
0.04	18.85	1.84
0.04	23.06	1.81
0.04	27.23	1.73
0.05	18.85	1.83
0.05	23.06	1.78
0.05	27.23	1.64

TABLE M5: INCREASE DUE TO WAVY FLOW

IMMERSION (m)	ANGULAR VELOCITY (rad/s)	k_{la} (h^{-1})	$k_{la(w)}$ (wavy flow) (h^{-1})	INCREASE (%)
0.03	18.85	0.64	1.82	184
0.03	23.06	0.64	1.82	184
0.03	27.23	0.67	1.77	164
0.04	18.85	0.64	1.84	188
0.04	23.06	0.72	1.81	151
0.04	27.23	0.72	1.73	140
0.05	18.85	0.52	1.83	252
0.05	23.06	0.52	1.78	212
0.05	27.23	0.57	1.64	188

TABLE M6: OVERALL MASS TRANSFER COEFFICIENTS FOR THE CFR [1]

ROTATIONAL SPEED (rad/s)	k_{la} (hr^{-1})
18.85	5.6
23.06	7.7
27.23	9.9

TABLE M7: COMPARISON OF WAVY FILM MASS TRANSFER COEFFICIENT TO THE OVERALL MASS TRANSFER COEFFICIENT.

ANGULAR VELOCITY (rad/s)	k_{la} (h^{-1})	$k_{la(w)}$ (h^{-1})	% $k_{la(w)}$ of k_{la}
18.85	5.6	1.83	32
23.06	7.7	1.78	23
27.23	9.9	1.64	16

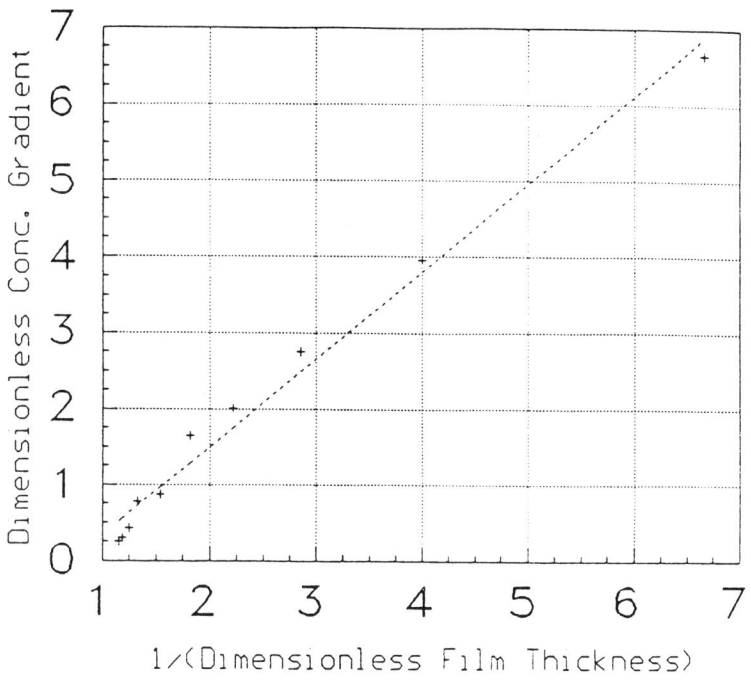

Figure 16 Plot of Concentration Gradient at Film Surface ($\Theta |_{\delta_+}$) vs $1/\delta^+$

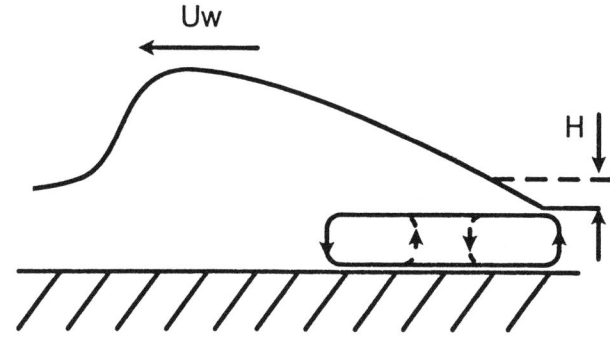

Figure 17: Circulating Eddy [10]

Semi-Direct Simulation of Flow in Turbulent Transition in a Vessel with Paddle Impeller

Yoichi Nagase, Kenji Nakamura
Chemical Engineering Department, Hiroshima University, Higashi, Hiroshima, Japan

Takahide Nouzawa
Vehicle Testing & Research Group, Mazda Motor Corporation, Hiroshima, Japan

Takeo Shiojima
Engineering Department, Idemitsu Petrochemical Corporation, Ichihara, Japan

Direct numerical simulation (DNS) of the flow field development from rest to the steady state flow at a Reynolds number of 2075 is examined. To reduce the computation cost, the simulation is limited to the upper half of the vessel without baffles. A flat blade turbine is located in the center of the vessel. The computation is unsteady. However, flow rate of shear and vorticity profile within the impeller region are calculated. These can hardly be observed by means of the flow measurement.

There have been published several papers which attempted to stimulate numerically the turbulent flow in the mixers. The first attempt was perhaps the work of DeSouza and Pike [1] who assumed a simplified model equation to predict the flow field in the bulk of the vessel. Work of Fort et al [2] is also available. Most papers (Harvy and Greaves [3], Placek and Tavlarides [4], Middleton et al [5], Ranade and Joshi [6], Ju et al [7]) applied the k-ε or modified k-ε models to the Reynolds stress equations to examine the flow field in the mixers. A review for the papers before 1990 was given by Ranade and Joshi [6]. Bakker and Akker [8] attempted to predict more complicate flow in the gas-liquid mixing using the algebraic stress model.

All of above authors modeled the time averaged properties of the turbulent flow in the bulk flow field in the vessels. The Reynolds stress is in general the sum of all the contributions of the stresses which are resulted from the interaction between turbulent velocity fluctuations with different wave numbers. Therefore, if the turbulent flow field is well developed and if the turbulence structure is more complicate than that in the well know turbulent field, then the usefulness of the k-ε model to predict the Reynolds stress is reduced. Turbulent flow in the mixer is one of complicated turbulence which have been pointed out by these authors [9,10,11,12].

An objective of this work is to analyze flow field in which measurement is hard to obtain, near field of the impeller blade for example. We applied the Direct Numerical Simulation (DNS) technique to simulate the flow near the rotating blade in the transition Reynolds number. A rather easy scheme has used. The calculations were limited to the upper half of the unbaffled vessel. This minimized the memory requirements of the computer. Still interesting results were found.

NUMERICAL METHOD

So called "Kawamura-Kuwahara (KK) scheme" was applied here [13]. The main procedure was to apply the modified Marker and Cell (MAC) method for the pressure and modified third accurate difference for the convection term. The former was obtained after applying the divergence to the Navier-Stokes equation. We get:

$$\nabla^2 p = -\nabla \cdot \{(u \cdot \nabla)u\} - \frac{\nabla \cdot u^{n+1} - \nabla \cdot u^n}{\Delta t} \quad (1)$$

Keeping $\nabla \cdot u^n$ in the second term and approximating the first term of right hand side, Equation (1) is rearranged to

$$\nabla^2 p^{n+1} = -(\nabla \cdot u^n)^2 + \frac{\nabla \cdot u^n}{\Delta t} \quad (2)$$

Equation (2) gives pressure at one time step up from the right hand side.

The time-stepping for the velocity u^{n+1} is obtained from Equation (3), after linearizing for convective term of N equation

$$\frac{u^{n+1} - u^n}{\Delta t} + (u^n \cdot \nabla)u^{n+1} = -\nabla p + \frac{1}{Re}\nabla^2 u^{n+1} \quad (3)$$

The upwind finite difference for the convective term

the KK scheme is

$$f\frac{\partial u}{\partial x}\bigg)_i = f_i \frac{-u_{i+2} + 8(u_{i+1} - u_{i-1}) + u_{i-2}}{12h}$$

$$+ |f_i| \frac{u_{i+2} - 4u_{i+1} + 6u_i - 4u_{i-1} + u_{i-2}}{4h} \quad (4)$$

where h is special difference.
Taylor expand of the right hand side of above equation gives

$$\text{right hand side} = \left(f\frac{\partial u}{\partial x}\right)_i + \frac{1}{4}h^3|f_i|\frac{\partial^4 u}{\partial x^4} + 0(h^4)$$

Therefore, Equation (4) is one of the third order accuracy. Aliasing error comes from the residual term of above equation and is negligible for this transition calculation. The central difference applies for the other terms in Equation (3).

$$\left(\frac{\partial u}{\partial x}\right)_i = \frac{u_{i+1} - u_{i-1}}{2h}$$

$$\left(\frac{\partial^2 u}{\partial x^2}\right)_i = \frac{u_{i+1} - 2u_i + u_{i-1}}{h^2} \quad (5)$$

Substituting Equation (4) and (5) and solution p^{n+1} of equation (2) into Equation (3), the u^{n+1} is calculated applying the Successive Over-Relaxation (SOR) method.

MIXER FOR SIMULATION AND GRID FORMATION

The impeller for the DNS was selected to the most simple one, a flat blade in the center of a cylindrical vessel without baffle as shown in Figure 1. Dimensions are also given in Figure 1. The cylindrical coordinate was used for the physical plane. Algebraic grid generation method given in Reference[14] was applied to give a dense mesh around the impeller blade and near the wall and shaft.
Equal angle of grids with $\Delta\theta = \pi/60$ to the θ direction were first given (main grid) for the full space, 2π radian, and then dense, near the impeller blade, to coarse ($=\pi/60$) grids (sub-grid) were given within twelve of the main grid distance $\pi/5$.
Grid is shown in Figure 2(a) and (b). The impeller blade is replaced with a wedge shape formed by the surface of the smallest subgrid, of which angle $\Delta\theta = 2.122$ degree. This is equivalent to 1.85 mm of the blade thickness at the impeller tip. Smallest distances of the boundary grid were 0.99 mm to the θ direction at the impeller tip, and 1.473 mm at the vessel wall. Similar grid distances about 1 to 2 mm were given to r- and z-direction within the impeller region.
Subgrid was fixed with the rotating impeller blade to give the new grid frame around the impeller. Initial velocities on the new grid were given by interpolation of the velocities at one-step before, $t=t^n$ and calculation is started for new time step. The impeller was started with a acceleration velocity 0.5 rad.s^{-2} until steady rotation speed of 0.2075 rps, which correspond to Reynolds number 2075, and then rotation was continued. Number of grid point including subgrid is 78x138x50=538200. Accuracy of calculation was 0.5% for pressure and 0.1% for velocities.
Boundary conditions are,
(i) on the shaft and the impeller blade:
$$u_r = u_z = 0, \quad u_\theta = r\omega$$
(ii) on the vessel wall:
$$u_r = u_\theta = u_z = 0$$
(iii) on the liquid surface and the mid plane in the vessel:
$$u_z = 0$$
The last condition for the mid plane are not recommended for the real flow simulation, but are inevitable to reduce the grid number. Calculation was performed by university computer server Convex C3240 together with end-processor C3210.

RESULT AND DISCUSSION

An illustration of the velocity field at an unsteady state that is at 4.5 revolution (23.3 s) after starting will be given here. Illustrations of two dimensional velocity vector distributions on r and z plane are given in Figure 3(a)~(d). From these figures, it is seen that flow field is still limited within and very near the impeller region. Inward and up flow in average is seen within $2r/D < 0.5$ as seen in Figure 3(a)(b). This upward flow turns toward outward, carry over just above the impeller and then returns to become down flow toward around the impeller tip region as shown in Figure 3(c)(d). Relatively large scale vortex appears during these small scale circulation, which is indicated (X) mark in Figure 3(c).
A peculiar circulation flow seen in Figure 3 is believed to come from that the u_θ just outside the impeller tip has little accelerated in this short period of rotation, therefore Coriolis's acceleration directing inward, and also come from the wedge shape of the blade as mentioned above, on which boundary velocity inclined inward even small angle. Flow development with further rotation of this artificial blade may be delayed than that of the real flow field mixed with the impeller of constant blade thickness.
Though flow field is somewhat peculiar, the local pressure p, rate of shear γ_{ij} and vorticity ω_i give reasonable distributions. Pressure is given in relative from average pressure over all the calculated space.
Figure 4(a)(b) give pressure distributions on the plane z/w=0.48 front (Figure 4(a)) and behind (Figure 4(b)) the blade respectively, on which most pressures are larger than those on the other plane at the corresponding (r,θ) points. It is seen from the figure that maximum pressure appears on midway of the blade in behind the zone of the blade, but maximum pressure appears at z/w=0.67 in

front zone.

Negative pressure appears near the shaft $2r/D < 0.30$ in the front zone and $2r/D < 0.26$ in the back zone (Negative pressure indicated arrow mark). The negative pressure is induced after flow direction changes inward and up.

A six components of γ_{ij}, $\gamma_{\theta\theta}$'s are largest in general than the other components. Figure 5(a)(b) show $\gamma_{\theta\theta}$ in the front zone, and in the back zone, respectively. It should be noted that most of $\gamma_{\theta\theta}$'s in the front and in the back zone except several point are plus or minus, respectively (opposite value in each zone indicated arrow mark). This means direct acceleration of the circumferential velocity, in the back zone is more. Fluid motion by the blade is reasonably effective in this acceleration period. In both zones, $\gamma_{\theta\theta}$'s are large near blade upper edge $z/w=1.0$ or 0.96 and then gradually decrease with z/w decrease to become 10^{-4} order on $z/w=0$.

Large minus values of the $\gamma_{\theta z}$ and γ_{rz} appear on the blade upper edge in both front and back zone of the blade as shown in Figure 6(a)(b) and Figure 7(a)(b), respectively. This is clearly due to edge effect of the blade. Both $\gamma_{\theta z}$ and γ_{rz} decrease as z decreases, to be order of 10^{-4} on the plane $z=0$. The $\gamma_{r\theta}$ is only appears on the blade tip corner as shown in Figure 8(a)(b). Ranges of γ_{rr} and γ_{zz} in the same zone as given above figures are $-0.80\sim 2.2$ and $-0.54\sim 11.4$, respectively.

Large variation of u_θ in z direction near upper edge and near the tip of the blade seen in Figure 3(c) and (d) is expected to be large value of ω_r and ω_z in corresponding regions. This is seen in Figure 9(a)(b) and Figure 10(a)(b), respectively. Every figure include the plane out of the impeller tip, $2r/D = 1.02$. Large ω_r's are seen on the upper edge of the blade in the back zone, but ω_r quickly decrease with z decreasing to be order 10^{-4}. On the other hand, ω_r is small in the front zone. In general, the rotating impeller along z-axis induces vorticity ω_z. With inducing more radial flow by the flat blade, vortex axis is turned the more towards the r-direction. Thus, the large ω_r component must be appeared in the wider zone (especially in the back zone) near the blade as the acceleration of fluid progresses. Figure 9 gives similar tendency even circulation flow profile is unusual. Range of ω_θ in the zone as Figure 9 is $-3.5\sim 1.1$.

Velocity profile after six rotation is similar but enlarge the streaming zone to those shown in this paper. Large value of most of the shear rate components and ω_r are ω_z appear on the plane $z/w=1$ or on the blade chip seem to be resulted from the geometrically sharp edges which are created artificially, and from boundary condition (iii).

Within these limitation, however, it is very interesting whether flow field is going to develop to the usual steady flow field, and how circulation flow profile, rate of shear and vorticity vary with this development.

LITERATURE CITED

1. DeSouza, A. and Pike, R.W., *Can. J. Chem. Eng.*, **50**, 15 (1972).
2. Fort, I., Obeid, A. and Brezina, V., *Czech. Chem. Commun.*, **47**, 226 (1982).
3. Harvey, P.S. and Greaves, M., *Trans. I Chem. E.*, **60**, 201 (1982).
4. Placek, J. and Tavlarides, L.L., 1985, *AIChEJ*, 31: 1113.
5. Middleton, J.C., Pierce, F. and Lynch, P.M., *Chem. Eng. Res. Des.*, **64**, 34 (1986).
6. Ranade, V.V. and Joshi, J.B., *Trans. I. Chem. E.*, **68**, 34 (1990).
7. Ju, S.Y., Mulvahill, T.M. and Pike, R.W., *Can. J. Chem. Eng.*, **68**, 3 (1990).
8. Bakker, A. and Van Den Akker, H.E.A., *Proc. 7th European Mixing Conf.* Part 1, 199 (1991).
9. Winardi, S. and Nagase, Y., *Chem. Eng. Commun.*, in press.
10. Winardi, S., Nakao, S. and Nagase, Y., *J. Chem. Eng. Japan*, **21**, 503 (1988).
11. Winardi, S., Kuwai, M. and Nagase, Y., *Mem. Fac. Eng. Hiroshima Univ.*, **10**, 37 (1990).
12. Hirofuji, Y., Kato, T. and Nagase, Y., *Bulletin Fac. Eng. Hiroshima Univ.*, **43**, 65 (1994) (Japanese).
13. Kuwahara, K., *J. Phys. Soc. Japan*, **40**, 877 (1985) (Japanese).
14. Fletcher, C.A.J., *Computational Techniques for Fluid Dynamics*, 2nd ed., Vol. 2, Springer-Verlag, Berlin, p.105 (1991).

CONCLUDING REMARK

Calculations have been employed the university computer. Calculation is completed at present around six rotation, net CPU time of about 540 hr. during last 6 months.

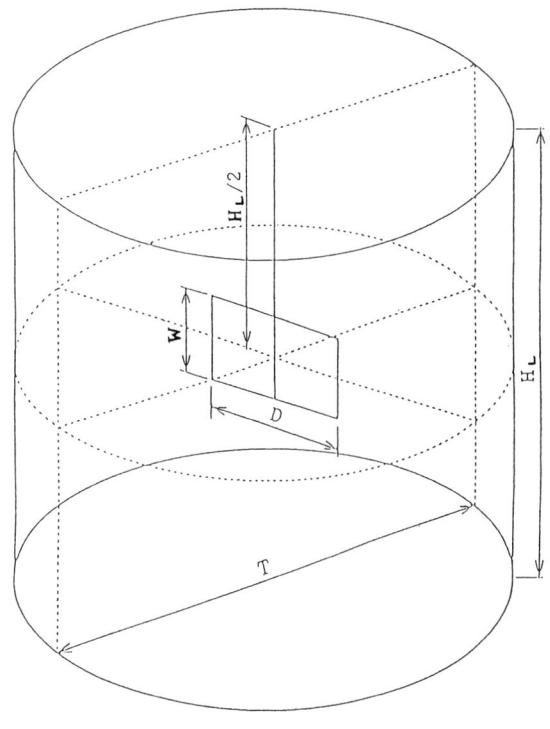

$D/T=1/3$, $H_L/T=1$, $W/D=1/2$, $T=300[mm]$

Figure 1 Geometry for the computation

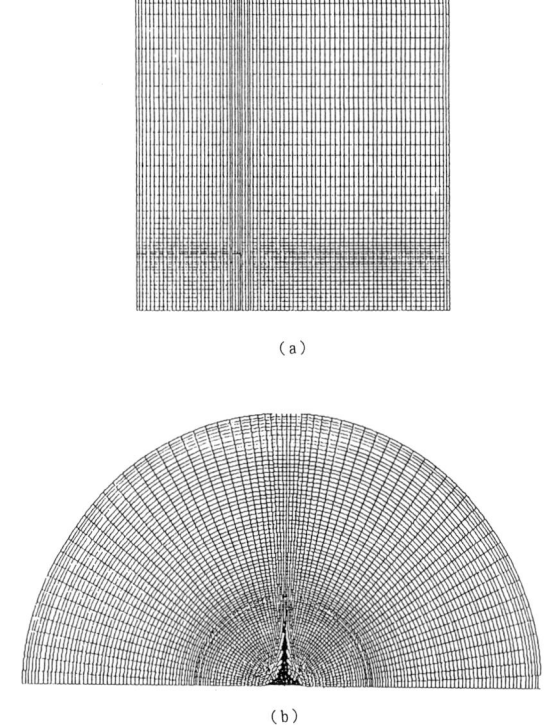

Figure 2 Grid frame when impeller blade locates at $\theta = \pi/2$

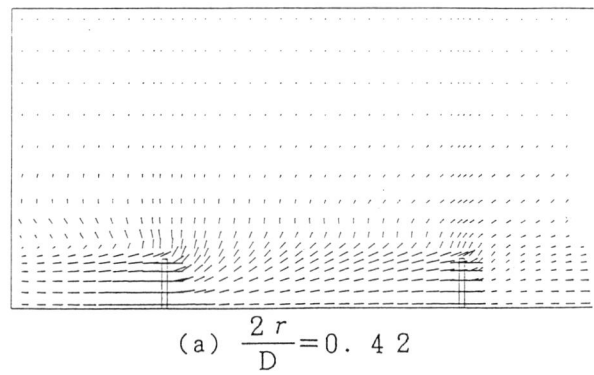

(a) $\dfrac{2r}{D}=0.42$

Figure 3 Illustrations of flow around the impeller

92 Industrial Mixing Fundamentals with Applications

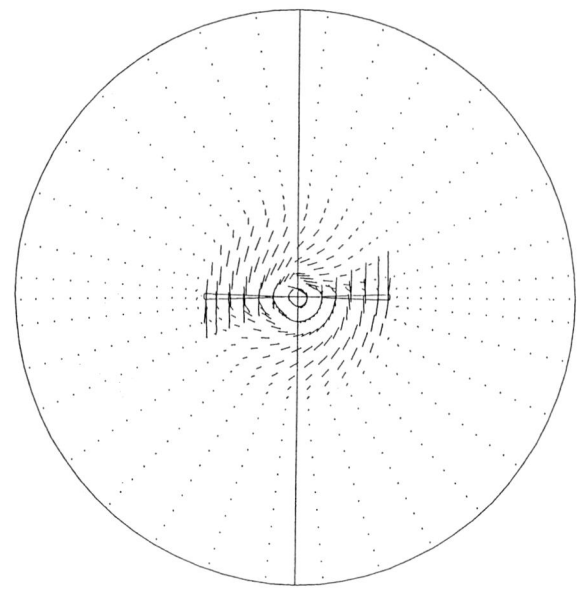

Figure 3 (b) $\frac{Z}{W}=0.55$

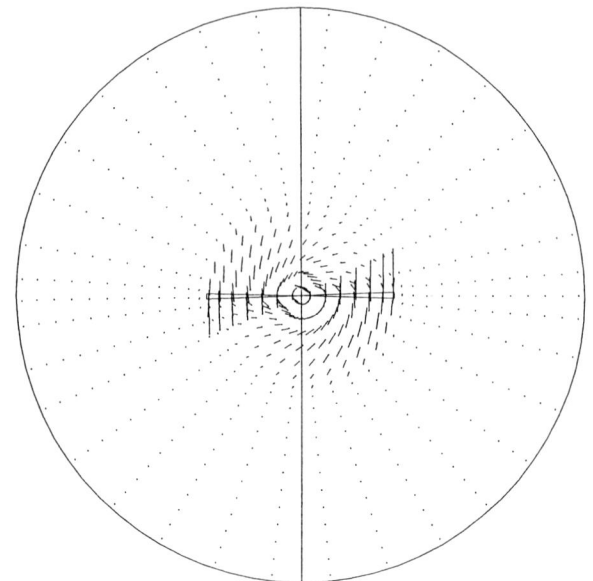

Figure 3 (d) $\frac{Z}{W}=1.0$

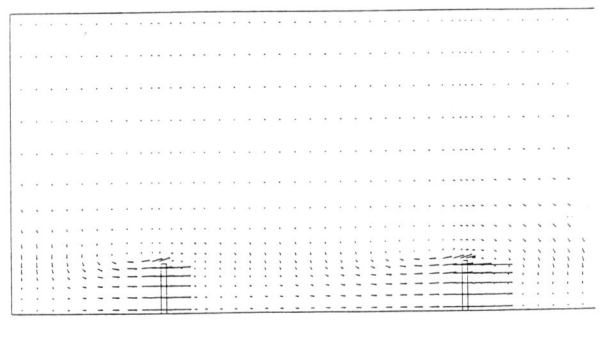

Figure 3 (c) $\frac{2r}{D}=1.0$

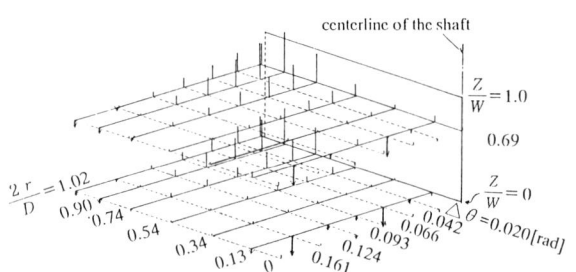

(a) front and near the blade - plus value shows upward

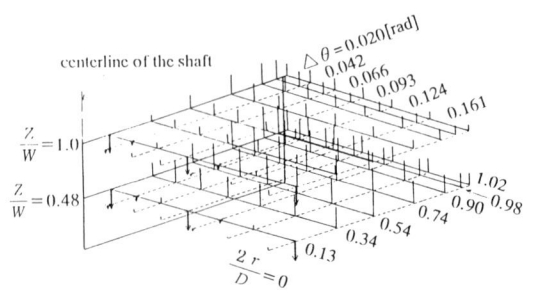

(b) behind and near the blade - plus value shows upward

Figure 4 static pressure distribution

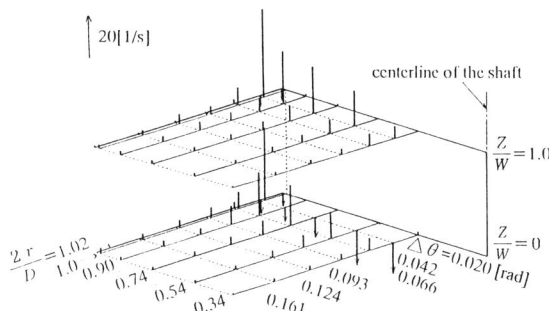

(a) front and near the blade - plus value shows upward

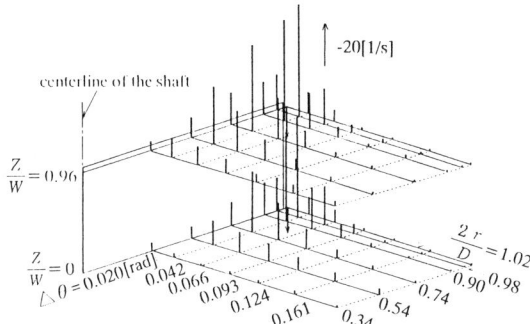

(b) behind and near the blade - minimum value shows upward

Figure 5 shear rate $\gamma_{\theta\theta}$ distribution

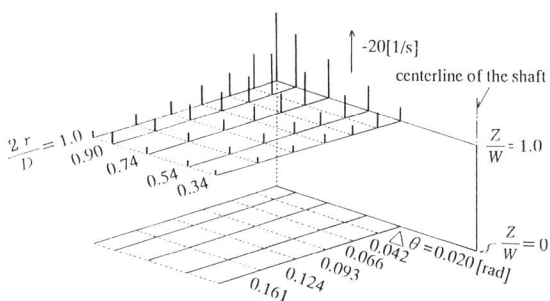

(a) front and near the blade - minimum value shows upward

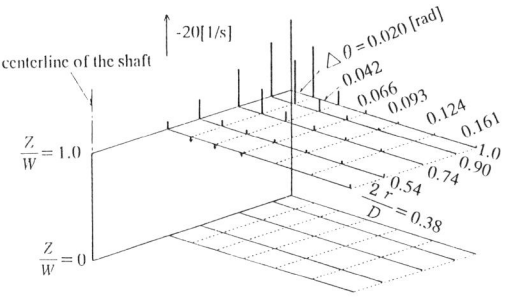

(b) behind and near the blade - minimum value shows upward

Figure 6 shear rate $\gamma_{\theta z}$ distribution

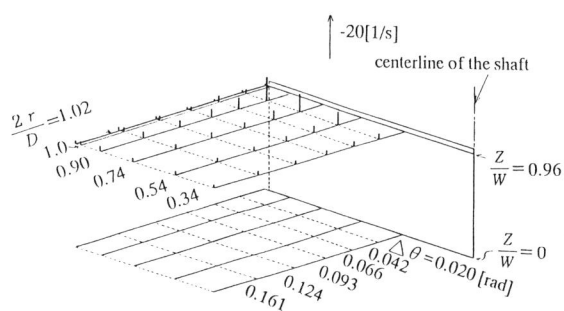

(a) front and near the blade - minimum value shows upward

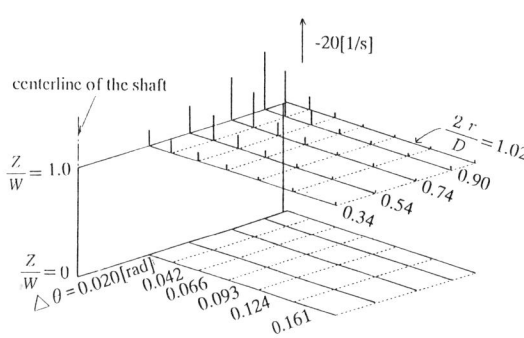

(b) behind and near the blade - minimum value shows upward

Figure 7 shear rate γ_{rz} distribution

94 Industrial Mixing Fundamentals with Applications   AIChE SYMPOSIUM SERIES

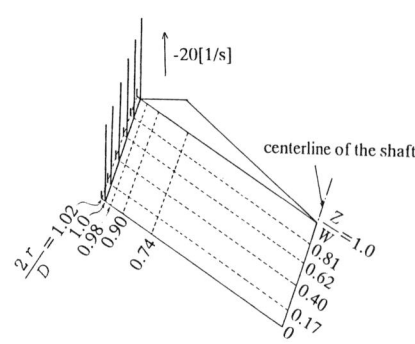

(a) front and near the blade - minimum value shows upward

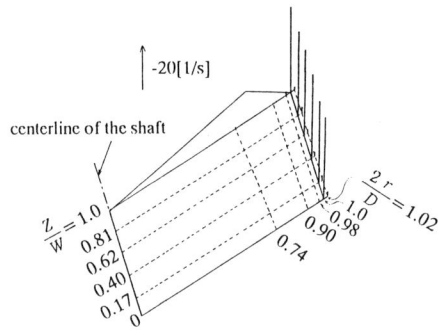

(b) behind and near the blade - minimum value shows upward

Figure 8 shear rate $\gamma_{r\theta}$ distribution

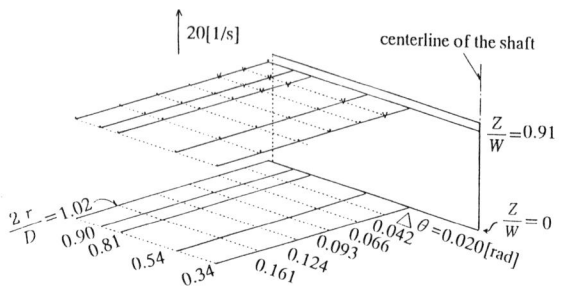

(a) front and near the blade - plus value shows upward

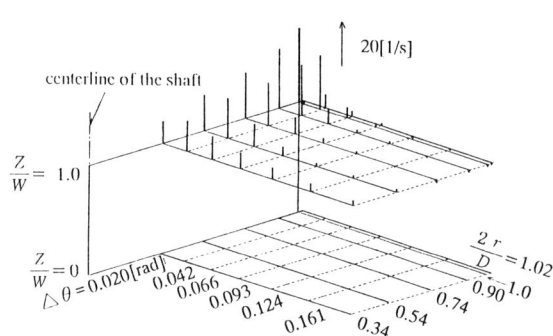

(b) behind and near the blade - plus value shows upward

Figure 9 shear rate ω_r distribution

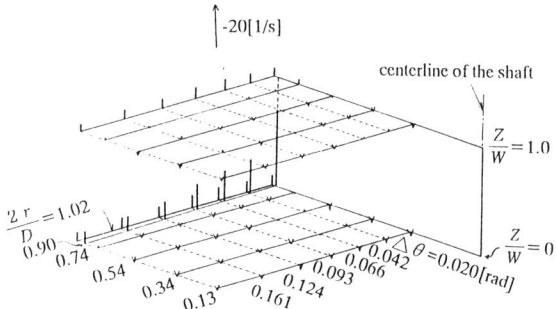

(a) front and near the blade - minimum value shows upward

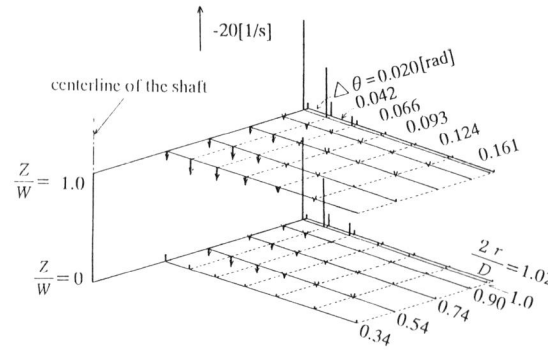

(b) behind and near the blade - minimum value shows upward

Figure 10 shear rate ω_z distribution

A New Scale-up Rule and Evaluation of Traditional Rules from a Viewpoint of Energy Spectrum Function

Kohei Ogawa and Chiaki Kuroda
Department of Chemical Engineering, Tokyo Institute of Technology, Tokyo, Japan

The most reliable scale-up rule is discussed based on energy spectrum function (ESF) curves which show the mechanism of turbulent flow field in stirred vessels. Based on the newly derived expression of one-dimensional ESF for wide wavenumber ranges, which is confirmed to be practically suitable, a new conception of scale-up is presented. If scale-up ratio is less than 27 in volume, the scale-up is completely successful by holding the impeller tip velocity constant. In this case, the ESF curve in the large stirred vessel is made to be the same as that in the small one. Even if scale-up ratio is set as more than 27 in volume, it is possible to succeed in scale-up if the power consumption per unit volume of vessel is held constant when small wavenumber range plays significant role in process. In this case, the ESF curves of the large and small stirred vessels overlap each other in the wavenumber range higher than that to which the Kolmogoroff spectrum law can apply.

One of major problems in industrial processes is successful scale-up of equipments under turbulent conditions from small to large size. Failure to scale-up properly stirred vessels, for example, remains a persistent problem in process industries. However, many methods of scale-up have been proposed by Paul and Treybal [1], Oldshue [2], Paul [3], Rice [4], Dickey [5], Leng [6], and so on, on the premise that all factors about the equipment geometry are similar between the small and the large equipment. No practical method, which has theoretical consideration, has been established. Of course, it is well understood that, if the turbulent phenomena in the large equipment are the same as those in the small one, the scale-up will be successful. However, there has been no theoretical consideration to satisfy the basically required conditions for turbulent phenomena which govern the process in different equipments. On the other hand, it has been said, if processes require completely constant turbulent phenomena, conditions of geometric similarity need to be relaxed. Ogawa et al. [7,8] and Ogawa [9] have studied the energy spectrum function(ESF) as one of representative factors which can express the mechanism of turbulent flow field in order to solve the problem.

The purpose of this paper is to propose a new way of scale-up of equipment under turbulent conditions by establishing the energy spectrum function (ESF) and to evaluate common scale-up rules based on the newly proposed conception.

Expression of ESF and scale-up Conception

Expression of ESF

For discussion about expression of ESF, first, the following are assumed for turbulent flow fields.

1. Turbulent flow field consists of m eddy groups: the basic/smallest eddy-group and its sequential subharmonic m-1 eddy-groups.
2. Eddy-group j contributes to the ESF curve which associates the maximum amount of information entropy under the condition that the average wavenumber K_j exists.
3. Weight factor of the turbulent kinetic energy of eddy-group j to total of that, u^2, is P_j. Summing over all eddies,

$$\sum_{j=1}^{m} P_j = 1 \tag{1}$$

Under above assumptions, the one-dimensional ESF curve is derived as follows.

$$E = \sum_{j=1}^{m} \frac{u^2 P_j}{K_j} \exp\left(-\frac{k}{K_j}\right) \tag{2}$$

Second, it is supposed that the following simple relationships exist among the average wavenumbers and the weight factors of each eddy-group, respectively.

$$\frac{K_j}{K_{j-1}} = \alpha, \quad \frac{P_j}{P_{j-1}} = \beta \tag{3}$$

Above Equation (2) is rewritten as follows by using above factors α and β.

$$E = \frac{u^2}{K_1 \sum_{j=1}^{m} \beta^{j-1}} \sum_{j=1}^{m} \left\{ \left(\frac{\beta}{\alpha}\right)^{j-1} \exp\left(-\frac{1}{\alpha^{j-1}} \frac{k}{K_1}\right) \right\} \tag{4}$$

By comparing the ESF curves expressed by Equation (4) under many combinations of α and β, the most suitable combination for them was found as

$$\alpha = 1/3, \quad \beta = 2 \quad (5)$$

Finally, substituting these values into Equation (4), a simple expression of one-dimensional ESF for wide wavenumber ranges is proposed.

$$\frac{EK_1}{u^2} = \frac{1}{\sum_{j=1}^{m} 2^{j-1}} \sum_{j=1}^{m} \{6^{j-1}\exp(-3^{j-1}\frac{k}{K_1})\} \quad (6)$$

It was confirmed by Ogawa et al.[7] that the ESF curve expressed by Equation (6) for each value of m appears to be a satisfactory approximation typical of those measured turbulent flows; pipe flow, jet flow, downstream of a grid and so on, and in the discharge flow in a stirred vessel with six-blade disk turbine type impeller as shown in **Figures 1 and 2**.

2. Scale-up Conception

As shown in **Figure 3**, in the case of fully developed turbulent pipe flows, Ogawa et al.[8] confirmed that there was a clear relationship between the pipe inner diameter and the number of eddy-groups m; a new subharmonic larger eddy-group appears at least when the pipe inner diameter becomes triple. In generally, this relationship between the size of vessel, D, and the critical size of vessel, D_i, can be written as follows.

$$m = 1; \quad 0 < D \leq D_1$$
$$m = i \geq 2; \quad D_{i-1} < D \leq D_i \quad (7)$$

Under the assumption that the critical size of vessel, D_i, is proportional to the average wavenumber K_i as

$$D_i = \gamma K_i^{-1} = \gamma 3^{i-1} K_1^{-1} \quad (8)$$

the above relational Equation (7) can be written as follows.

$$m = 1; \quad 0 < D \leq \gamma K_1^{-1}$$
$$m = i \geq 2; \quad \gamma K_{i-1}^{-1} = \gamma 3^{i-2} K_1^{-1}$$
$$< D \leq \gamma K_i^{-1} = \gamma 3^{i-1} K_1^{-1} \quad (9)$$

It is easy to estimate that the scale of the smallest eddy-group, which corresponds to K_1^{-1}, depends only on the fluid properties and has no relation to the size of equipment. Considering the fluid in the larger vessel is the same as that in the smaller vessel, the average wavenumber of the basic/smallest eddy-group, K_1, becomes the same as that in the small vessel. Therefore, when the turbulent kinematic energy is changed, only up and down movement of the ESF curve is expected in the figure of ESF vs. wavenumber. It is possibile to move the ESF curve of the large vessel up and down and to overlap that of the small vessel in the figure.

According to above considerations, the degree of overlap and the relative magnitude of the turbulent kinetic energy of the larger vessel, u_i^2, to that for smaller vessel become important. For the discussion, the value of exponent δ in the following relation becomes important.

$$u_i^2 = (\frac{D_i}{D_1})^\delta u_1^2 = (3^{i-1})^\delta u_1^2 \quad (10)$$

By changing the value of δ, the up and down movement of ESF curve, that is to say, the degree of the overlap of the ESF curves in the figure is investigated. Representative results are shown in **Figure 4**. In the figure, each ESF curve is expressed the relative to that for the case of m=1. The curve for m=i is the results when the size of large vessel is 3^{i-1} times of that of the case of m=1. If there is the wavenumber range where all curves are overlapped, the value of δ will be valuable regardless the value of scale-up ratio when the wavenumber range plays significant role for the process.

From these results it can be said that when the value of the exponent becomes 2/3, the ESF curves overlap each other in the wavenumber range higher than that to which the Kolmogoroff spectrum law can apply. As a result,

$$\frac{u_i^2}{D_i^{2/3}} = \frac{u_1^2}{D_1^{2/3}} \quad (11)$$

On the other hand, when the value of the exponent becomes -1, the ESF curves roughly overlap each other only in the lower wavenumber range. When the other values of δ are used, there is no wide region of overlap. Of course, if the size of the large vessel is less than triple of that of the small one, the ESF curves before and after scale-up become the same. In this case, the turbulent kinetic energy, u^2, should be held constant, whether the process is governed by any wavenumber.

Finally, the following scale-up method can be presented from the viewpoint of overlap of ESF curves.

1. Scale-up is successful if the scale-up ratio is less than 27 in volume by holding the turbulent kinetic energy constant.
2. If the scale-up ratio is more than 27 in volume, the scale-up is successful in two cases;
   $u^2/D^{2/3}$=constant when the higher wavenumber range plays significant role for the process, and
   $u^2 D$=constant when the lower wavenumber range plays significant role for the process.

PHISICAL MEANING OF SCALE-UP RULES OF STIRRED VESSEL

There are various rules recommended for scale-up similar stirred vessels under turbulent conditions. Some of the representative rules are shown in **Table 1**. In these rules, the impeller rotational speed can be replaced by the turbulent kinetic energy, e.g. Because, Ito et al [10] found experimentally that, $(u^2)^{1/2}$ is proportional to impeller tip speed ND.

In the following, the physical reliability of each rule in Table 1 is discussed from the viewpoint of the ESF expressed with Equation (6). ESF curves, when scale up is done by holding respective parameter constant, are the same as those shown in Figure 4. The value of the exponent in Equation (10) is selected under the consideration of the common scale-up rules in Table 1.

1. ND^0=constant

This condition corresponds to u^2D^{-2}= constant. The ESF curves when this parameter is held constant are shown in **Figure 4**. The curves do not touch or intersect. These results show that there is no physical meaning in this rule for scale-up from the viewpoint of ESF.

2. $ND^{2/3}$ =constant

This condition corresponds to $u^2D^{-2/3}$= constant. The ESF curves when this parameter is held constant are shown in **Figure 4**. In this case, holding the power consumption per unit volume constant is the same as holding $u^2D^{-2/3}$ constant. In the higher wavenumber range than that to which the Kolmogoroff spectrum law($E \propto k^{-5/3}$) can apply, all curves seem to overlap. This tendency is more complete in the case of m≥2. On the other hand, in the lower wavenumber ranges, there is no advantageous tendency for scale-up. These results show that there is a clear physical meaning in this rule for scale-up from the viewpoint of ESF if the turbulence phenomena in higher wavenumber range is significant, that is to say, for example microscale mixing is significant in the process.

The correspondence of the condition $u^2D^{-2/3}$= constant with the condition of constant of the power consumption per unit volume is established in only the case that the power number in the higher Reynolds number range takes a constant value. This usually occurs in the stirred vessels under the fully baffled condition but not necessarily in the other vessels.

3. ND=constant

This condition corresponds to u^2D^0= constant. The ESF curves when this parameter is held constant are shown in **Figure 4**. Every curve has only one cross point with another curve. From these results, it seems that there is no physical meaning in this rule for scale-up from the viewpoint of ESF.

However, this rule is valuable when the scale-up ratio is less than 27 in volume, because this rule means that the turbulent kinetic energy is held constant.

4. ND^2=constant

This condition corresponds to u^2D^2= constant. The ESF curves when this parameter is held constant are shown in **Figure 4**. Every curve has no cross point each other and they parallel each other in the lower wavenumber range. These results show that there is no physical meaning in the use of this parameter for scale-up.

5. $ND^{3/2}$=constant

This rule is out of Table 1. In the stirred vessel, this conditions means constant impeller discharge energy per unit discharge area and unit circulation time. This condition corresponds to u^2D=constant. This ESF curves when this parameter is held constant are shown in **Figure 4**. Though every curve has no cross point each other, they are very close each other in the lower wavenumber range. These results show that this rule will be valuable for scale-up if only the macroscale mixing is important in the process.

CONCLUSION

The physical meaning of common rules for scale-up vessels under turbulent conditions are evaluated from the viewpoint of the energy spectrum function which was proposed by making use of information entropy conception.

In order to succeed successfully at scale-up, the scale-up ratio should be set less than 27 in volume by holding the turbulent kinetic energy constant, that is to say, impeller tip velocity is held constant.

If the scale-up ratio is necessary to be set more than 27 in volume, the scale-up for the process in which the higher wavenumber range plays significant role is successful when $u^2/D^{2/3}$ is held constant, that is to say, the power consumption per unit volume of vessel is held constant. Additionally, the scale-up for the process in which the lower wavenumber range plays a significant role is successful when u^2D is held constant.

NOTATION

D = representative size of equipment [m]
D_1 = representative size of the basic equipment [m]
D_i = critical representative size of equipment for the number of eddy-group m [m]
E = one-dimensional energy spectrum [m^3/s^2]
K_1 = average wavenumber in case of m=1 [1/m]
K_j = average wavenumber of eddy-group j [1/m]
k = wavenumber [1/m]
m = number of eddy-group [-]
N = impeller rotational speed [1/s]
P_j = weight factor of turbulent kinetic energy of eddy-group j [-]
u^2 = turbulent kinetic energy [m^2/ s^2]
u_1^2 = critical turbulent kinetic energy corresponding to D_1 [m^2/s^2]
u_i^2 = critical turbulent kinetic energy corresponding to D_i [m^2/s^2]
α = ratio of average wavenumber [-]
β = ratio of weight factor of turbulent kinetic energy [-]
γ = proportional coefficient [-]
δ = exponent [-]

LITERATURE CITED

1. Paul. E. L. and R. E. Treybal, "Mixing and Product Distribution for a Liquid-Phase, Second-Order, Competitive-Consecutive Reaction", *AIChE J.*, **17**, 718 (1971).

2. Oldshue, J. Y. "Fluid Mixing Technology"., p.197, McGraw-Hill, New York, USA (1983).

3. Paul. E. L., "Design of Reaction Systems for Specialty Organic Chemicals", *Chem. Eng. Sci.*, **43**, 1773 (1988).

4. Rice, R. W. and R. E. Band, "The Role of Micromixing in the Scale-Up of Geometric Similar Batch Reactors", *AIChE J.*, **36**, 293 (1990).

5. Dickey, D. S., "Succeed at Stirred at-Tank-Reactor Design", *Chem. Eng. Prog.*, December, 22 (1991).

6. Leng,D. E., "Succeed at Scale Up",*Chem. Eng. Prog.*, June, 23 (1991)

7. Ogawa, K., C. Kuroda and S. Yoshikawa, "An expression of Energy Spectrum Function for Wide Wavenumber Ranges", *J. Chem. Eng. Japan*, **18**, 544 (1985).

8. Ogawa, K., C. Kuroda and S. Yoshikawa ., "A Method of Scaling Up Equipment from the Viewpoint of Energy Spectrum Function", *J. Chem. Eng. Japan*, **19**, 345 (1986).

9. Ogawa, K., "Energy Spectrum Function of Eddy Group Gathered Together as a Model of Turbulence Structure", *Int. J. Eng. Fluid Mechanics*, 1235 (1988).

10. Ito,S., K. Ogawa and N. Yoshida, "Turbulence in Impeller Stream in a Stirred Vessel", *J.Chem. Eng. Japan*, **8**, 206 (1975).

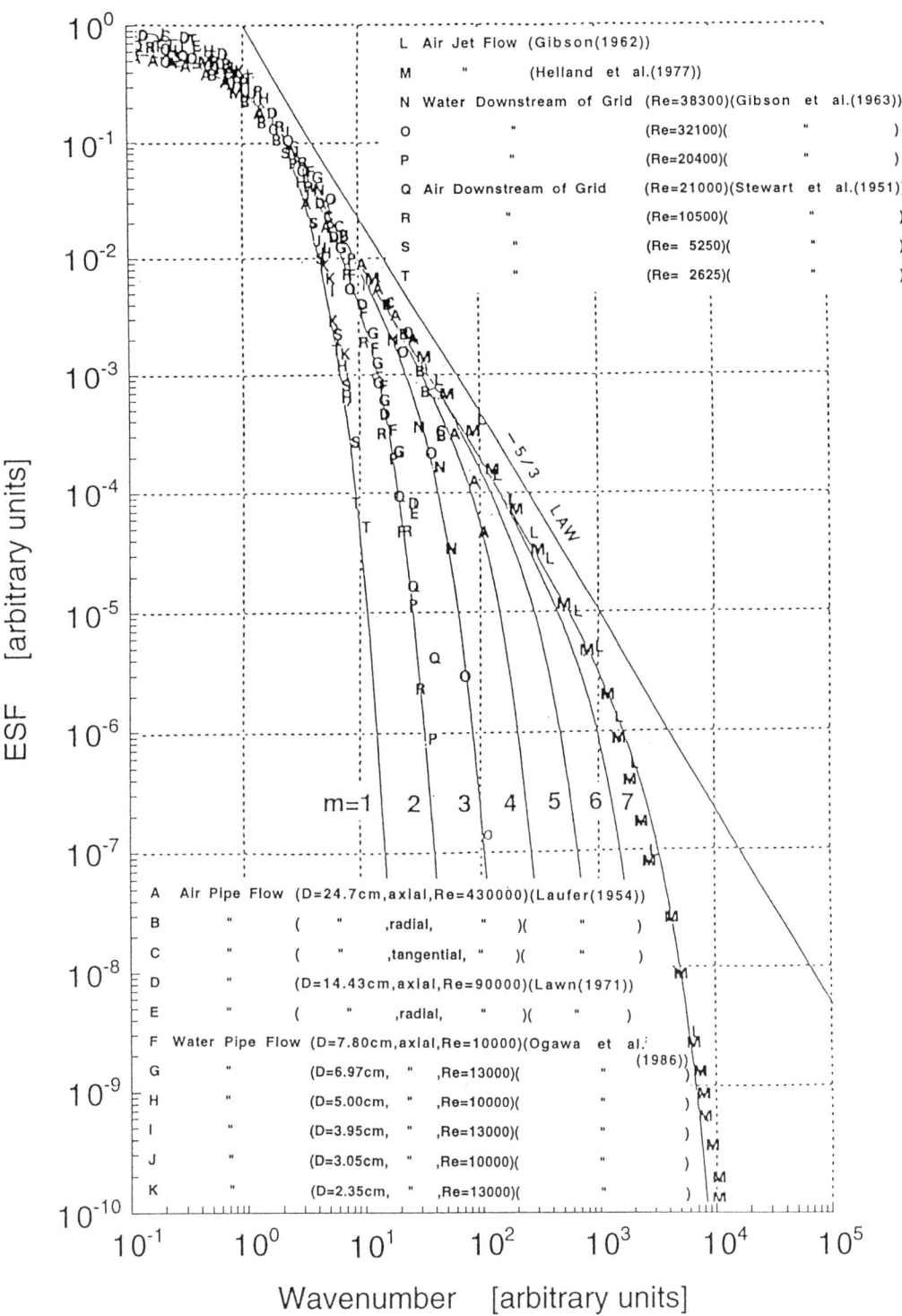

Figure 1   ESF curves measured in typical turbulent flows.

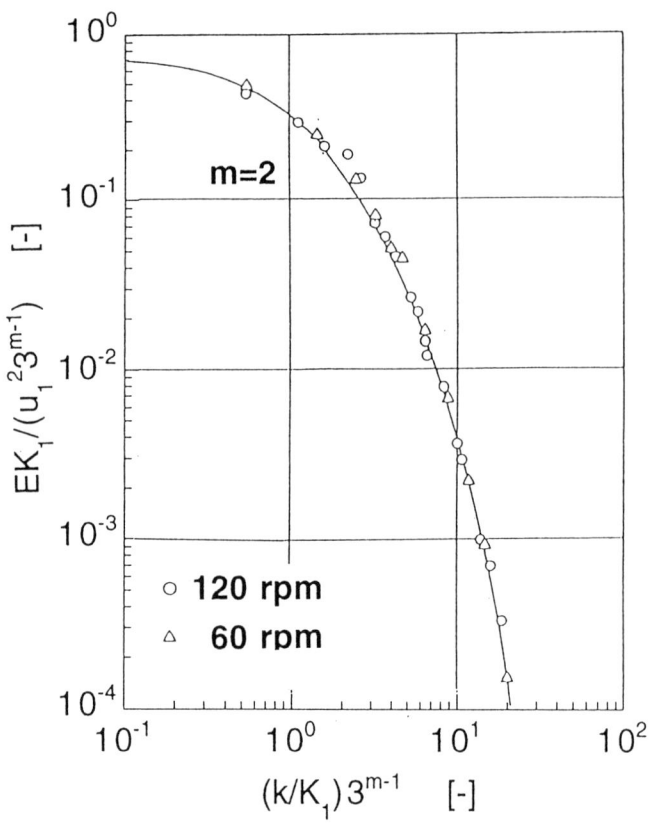

Figure 2 ESF curve in a stirred vessel.

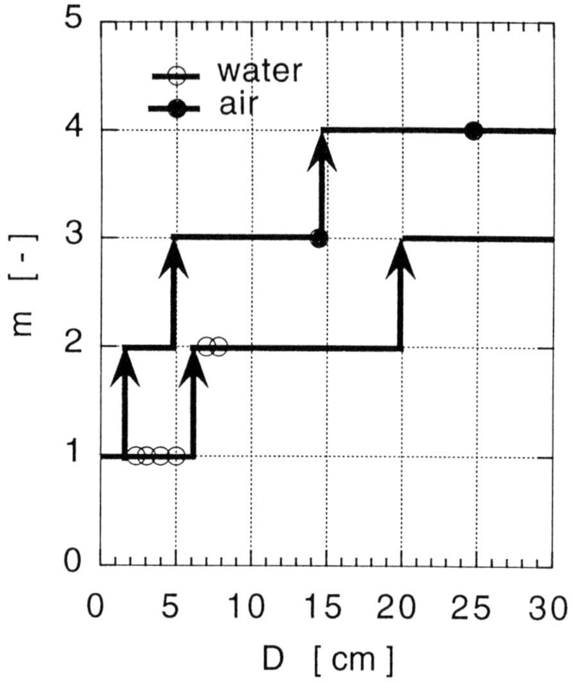

Figure 3 Relationship between m and pipe inner diameter

Table 1 Rules for scale-up similar vessels under turbulent conditions.

ND^X=const. Value of X	u^2D^Y=const. Value of Y	Rules	Processes
0	-2	Const. impeller revolutional speed Const. circulation time Const. impeller discharge flow rate per unit vessel volume	Fast reaction
2/3	-2/3	Const. (power) dissipation energy per unit vessel volume Const. impeller discharge flow energy	Turbulent dispersion Gas-liquid operation Reaction requiring microscale mixing
1	0	Const. impeller tip velocity Const. torque per unit vessel volume	
2	2	Const. Reynolds number Const. impeller discharge flow momemtum Const. torque per unit discharge flow rate	

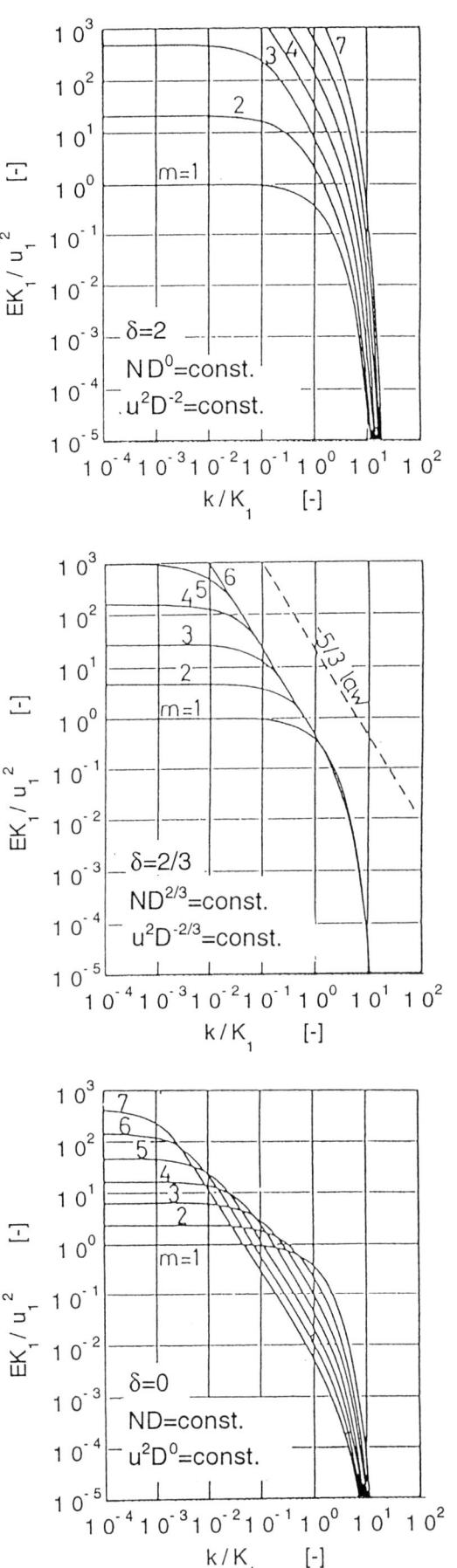
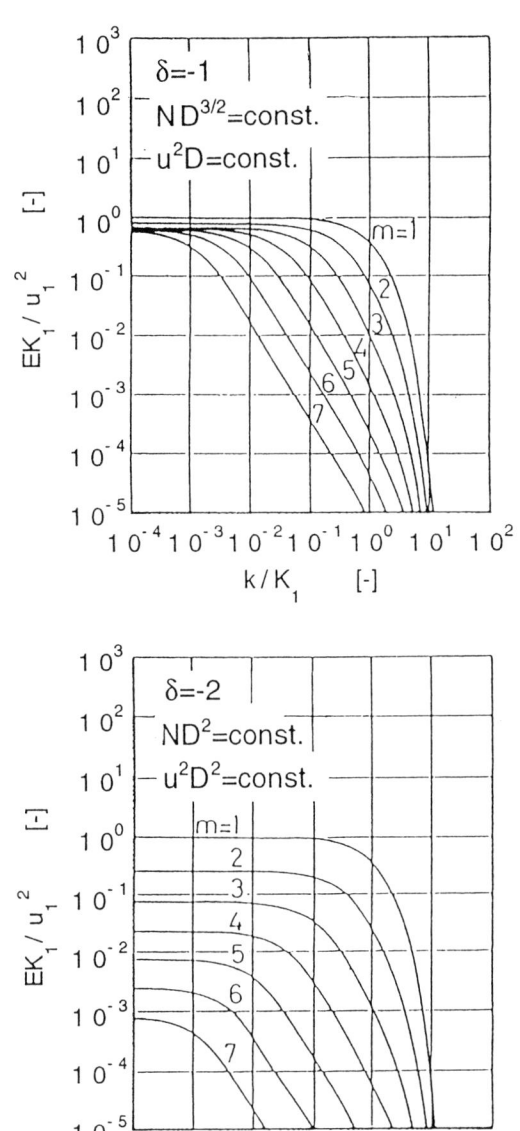

Figure 4 ESF curves in the case of ND^X=constant (u^2D^Y=constant)

An Experimental (LDA) and Numerical Study of the Turbulent Flow Behavior in the Near Wall and Bottom Regions in an Axially Stirred Vessel

Claes Sturesson, Hans Theliander and Anders Rasmuson
Department of Chemical Engineering Design, Chalmers University of Technology
S-41296 Gothenburg, Sweden

This study includes experimental and numerical characterization of the velocity field in the near bottom and wall regions of an axially stirred tank. The information is essential for a better understanding of the lifting mechanism of solid particles in the complete suspension process. The experimental investigation included single component LDA-measurements for three different impeller speeds, all in the turbulent regime of mixing, in a plane midway between the baffles. The mean and fluctuating radial and tangential velocity components in the boundary layer at the bottom of the tank were measured. The mean axial component of velocity close to the wall is reported for several different heights above the vessel bottom. Numerical calculations were performed using the standard k-ε turbulence model. A low Reynolds number k-ε turbulence model was compared to using wall functions in the near wall and bottom regions.

It was found that despite the presence of the solid boundary the mean and fluctuating velocities scale with impeller tip speed for the three impeller speeds studied. A linear mean velocity profile in the viscous sublayer and a maximum value of the fluctuating radial and tangential velocities close to the bottom of the tank were found. The development of the mean axial flow field along the vertical wall was characterized. Numerically obtained results showed qualitative agreement with experimental data.

A detailed characterization of the flow pattern near the wall and bottom regions in a stirred vessel is necessary in order to gain a better understanding of the mechanisms responsible for suspension of solid particles in a stirred vessel. The axial component of velocity at the tank wall is of primary interest since this component can be expected to contribute mostly to the lifting process of particles [1]. In previous investigations Pitot-tubes have been used, e.g. Fort et al. [2] and Placek et al. [3], for velocity measurements near the vessel walls. Near wall and bottom measurements in a stirred tank using the LDA-technique have to the authors knowledge not previously been reported in the literature. However, near wall LDA-measurements in channel flow systems (e.g. Rosén and Trägårdh [4] and Johansson [5]) have been performed.

The present study includes experimental and numerical characterization of the velocity field at selected locations in the near bottom and wall regions. The LDA-measurements were conducted using a DANTEC FiberFlow system with FFT signal processing to extract the velocity information. Beam expansion was used to minimize the size of the measuring control-volume. Impeller boundary conditions for the numerical calculations were obtained as whole-cycle-ensemble averages in a plane just below the impeller. It has been found in previous studies (e.g. Sturesson and Rasmuson [6]) that the turbulent flow field in the bulk region of an axially stirred tank predicted using the standard k-ε model correlates well with experimental LDA-data. Calculations in the present study were performed using a low Reynolds number k-ε turbulence model, which is a modification of the standard k-ε turbulence model, to allow calculations of turbulent flows at low Reynolds numbers. The numerically obtained single phase fluid flow near the base and walls of the vessel have been used to correlate experimental LDA-measurements. Numerical results were obtained using the commercial code FLOW3D (CFDS, Harwell Laboratories Version 3.2.1, 1993).

SYSTEM INVESTIGATED

Experimental laser Doppler anemometry (LDA) measurements and numerical calculations were performed for a cylindrical flat bottomed tank with four baffles equally spaced around the perimeter. The cylindrical vessel was made of plexiglass and filled with tap water. The geometry of the mixing system is depicted in Figure 1.

The vessel diameter (T) and the height of liquid above the bottom (H) were 0.289 m. A pitched blade impeller, with four blades inclined at an angle of 45°, pumping liquid downwards towards the bottom, was used. The impeller diameter (D) was 1/3 of the tank diameter and the ratio of the height of the impeller blade to the impeller diameter was 1/5. The projected impeller blade height was thus $0.2\sqrt{2}/2$ times the impeller diameter. The blade thickness was 1.5 mm and the hub diameter was 14 mm. The width of the non-transparent baffles (B) were 1/10 of the tank diameter and extended from the vessel bottom to the liquid surface. Experimental LDA measurements and numerical calculations were conducted for an impeller clearance above the bottom (C) of 1/3 of the vessel diameter. All experiments and numerical calculations were performed in the fully turbulent regime ($N_{Re} \geq 10^4$). LDA measurements were carried out at a plane midway between the baffles and at several different heights in the tank.

EXPERIMENTAL

The LDA set-up

Experimental measurements were conducted using a DANTEC FiberFlow system (Series 60X) and a Spectra-Physics (Model 2060A-64) water-cooled Ar-ion laser with a maximum output of 4W. A signal processor of the type DANTEC 57N10 Burst Spectrum Analyzer (BSA) was used to extract the velocity information from the Doppler signal using the FFT-technique. In order to resolve directional ambiguity a frequency shift of 40 MHz was imposed on one of the incoming beams using a Bragg cell. A fiberoptic probe with a focal length of 310 mm was used and included a receiving module, since the LDA-system was operated in the backscatter mode. A beam expander, with an expansion ratio of 1.94, was attached to the probe. The beam expander was used to reduce the size of the measuring volume and consisted of a concave lens which was placed before the convex lens which was used to bring the light in focus. A computer of the type Compaq ProLinea 4/33 automatically controled data acquisition, traversing and data manipulation using DANTEC BurstSize software. The fiberoptic probe was positioned on a DANTEC lightweight traversing system to allow movements in all three directions. The automatically controlled traversing mechanism allowed a repeatability of 40 μm and a resolution of 12.5 μm. The dual beam LDA-system used was operated in the single component mode and is illustrated in Figure 2.

The cylindrical vessel was surrounded by an outer square tank to minimize the effect of vessel curvature on the intersecting beams. The gap between the square glass container and the cylindrical vessel was filled with tap water in order to reduce errors when positioning the measuring control-volume. Silicone seeding particles with a mean diameter of 2 μm were added to the fluid to increase the data generation rate.

The distance between the incident beams was approximately 70 mm and formed an angle of $\alpha = 13.0°$ in air. The characteristics of the measuring control volume is shown in Table 1.

Measurement procedure

Measurements of the mean and root mean square (RMS) velocities at selected sampling locations were carried out to obtain information about the flow in the neighborhood of the walls and bottom of the axially stirred tank. The radial and tangential components of velocity (parallel to the vessel bottom) were measured from 50 μm to 5 mm from the vessel bottom at a radial distance of 70 mm from the centerline. This radial position was chosen since the radial velocity component could not be measured at the outer part of the vessel because of the presence of the non-transparent baffle and because of the effect of the strong curvature of the wall. This problem could be reduced by matching the refractive index of the fluid in the square tank. However, such matching of refractive index has not been attempted in this study. The axial velocity was measured at 5, 10, 20, and 30 mm above the vessel bottom and at radial distances ranging from 0.4 to 5 mm from the vessel wall. Measurements closer to the vertical wall was not feasible using the current optical arrangement because of the strong reflections from the vessel wall.

Measurements of the radial velocity were performed in a plane perpendicular to the optical axis of the incoming beams. The axial and tangential velocity components were measured in the plane passing through the optical axis of the beams. However, due to the difference in optical pathways in air and water a correction was required as described by Sturesson and Rasmuson [6]. The radial and tangential components of velocity were measured using the beams of wavelength 488 nm (blue) while the axial component was measured using the beams of wavelength 514.5 nm (green).

The effect of the impeller speed was studied using impeller rotational speeds of 3, 6, 9 rev/s. These impeller speeds corresponds to Reynolds numbers of mixing of 2.8×10^4, 5.6×10^4 and 8.4×10^4 respectively, which ensures the turbulent regime of mixing. In this study a sampling population of approximately 10000 at each measuring location was used to determine the mean and fluctuating velocities. However, the random error (precision) introduced by the finite number of samples was not investigated in detail. The reproducibility was determined from these measurements. The average validation of the samples was approximately 90 %. The data rate depended strongly on the sampling location and the impeller rotational speed. In this study the accuracy of the measurements of the mean velocity was less than ± 10 % and for the fluctuating velocity less than ± 5 % as determined from the measurements themselves. The given values are averages for all three impeller speeds.

The zero reference location of the measuring control volume with respect to the solid surface was determined by positioning the center of the volume at the surface of the wall (see Figure 3). This was achieved with an accuracy of approximately 0.1 mm at the wall and 0.01 mm at the vessel bottom by monitoring the current from the photomultiplier and adjusting the position of the measuring volume for maximum current.

It was found that measurements conducted when having the control-volume partially immersed in the vessel wall was disturbed because of the reflected light from the plexiglass wall. This fact gave rise to a broadened velocity distribution as exhibited on velocity histograms. Near bottom measurements of the radial and tangential velocity components could be performed at a distance of 50 μm from the solid surface with no influence of noise or reflections. At this location the edge of the measuring control volume was approximately 10 μm from the bottom surface. The velocity at a distance of 50 μm from the bottom surface showed a typical distribution as monitored on a histogram.

During the measurements of the axial component the control volume length was used which was less preferential with respect to spatial resolution. In addition, the influence of reflections were more significant when measuring close to the vertical tank wall compared to near bottom measurements. In order to obtain acceptable signal quality the first measuring position was not allowed closer than about 0.4 mm from the vertical wall. In this case the measuring control-volume was not immersed in the wall. However, despite that the first measuring location was 0.4 mm from the wall it was found that the axial velocity was somewhat disturbed because of the added noise from the light reflected from the vertical vessel wall as exhibited on a velocity histogram. The axial fluctuating velocities are therefore not reported while the mean values were judged to be less affected by the extraneous light.

THEORETICAL

The inadequacy of turbulent transport models to describe the influence of viscosity near the wall and the steep gradient of the calculated variables near the wall necessitate special treatment for computational nodes close to the wall. The wall function method and the low Reynolds number modeling method are the two main methods which can be used to account for the wall-proximity region when computing turbulent flows. Wall functions relate the velocity and turbulence parameters at the first computational node adjacent to the solid surface mainly to the friction velocity and are based on the assumption on the validity of local equilibrium of turbulence [7]. Wall functions have been widely used because of computational economy since an excessive amount of computational nodes in the near wall region is avoided. However, since wall functions are unsuited for the viscosity affected near wall region (the assumption of a balance of production and dissipation is not valid close to the wall) the first grid node adjacent to the wall should not be closer to the wall than approximately $y^+ = 30$.

For making the turbulence model applicable to the viscosity affected near wall region a low Reynolds number model can be applied. This model use modified turbulence constants and functions which are included in the turbulent transport equations for appropriate representation of the near wall region. When using the low Reynolds number models the transport equations are integrated to the wall through the viscous sublayer and thus a fine grid close to the solid surface is required. The standard no-slip conditions with zero values for k and ε are used. Thus, detailed knowledge of the velocity field close to solid surfaces can be obtained using the low Reynolds number model.

In the present study the k-ε turbulence model with the low Reynolds number near wall model was implemented for calculations using fine grid distributions near the bottom and wall of the tank. In the two equation k-ε turbulence model the Reynolds stresses are represented in analogy with molecular viscous stress using the Boussinesq eddy viscosity concept:

$$\rho\overline{u_i u_j} = -\mu_t\left(\frac{\partial \overline{U}_i}{\partial x_j} + \frac{\partial \overline{U}_j}{\partial x_i}\right) + \frac{2}{3}\delta_{ij}\rho k \quad (1)$$

The governing equations for the k-ε model have been prescribed previously by e.g. Sturesson et al. [8]. The low Reynolds number k-ε model is a modification of the standard k-ε turbulence model to allow calculations of turbulent flows at low Reynolds numbers. The model involves a damping of the eddy viscosity when the local turbulent Reynolds number is low, a modified definition of ε so that it goes to zero at walls and modifications of the source terms in the turbulent transport equations. For steady-state incompressible flow the equations describing the turbulence model become (FLOW3D, 1993):

$$\mu_t = C_\mu f_\mu \rho \frac{k^2}{\varepsilon} \quad (2)$$

$$\frac{\partial}{\partial x_j}(\rho \overline{U}_j k) = \frac{\partial}{\partial x_j}\left(\left(\mu + \frac{\mu_t}{\sigma_k}\right)\frac{\partial k}{\partial x_j}\right) + P_k - \rho\varepsilon - D \quad (3)$$

$$\frac{\partial}{\partial x_j}(\rho \overline{U}_j \varepsilon) =$$

$$= \frac{\partial}{\partial x_j}\left(\left(\mu + \frac{\mu_t}{\sigma_\varepsilon}\right)\frac{\partial \varepsilon}{\partial x_j}\right) + (C_{\varepsilon 1}P_k - C_{\varepsilon 2}\rho f_2\varepsilon)\left(\frac{\varepsilon}{k}\right) + E \quad (4)$$

The production term, P_k, which is the rate of generation of turbulence and the model constants are modeled according to the default expressions given in the FLOW3D manual.

The functions f_μ, f_2, D and E are defined by:

$$f_\mu = \exp\left(-\frac{3.4}{\left(1 + \frac{R_T}{50}\right)^2}\right) \quad (5)$$

$$f_2 = 1 - 0.3\exp(-R_T^2) \quad (6)$$

$$D = 2\mu\left(\frac{\partial k^{1/2}}{\partial x_j}\right)^2 \quad (7)$$

$$E = 2\frac{\mu\mu_t}{\rho}\left(\frac{\partial U_i^2}{\partial^2 x_j}\right)^2 \qquad (8)$$

where the local turbulent Reynolds number is defined by:

$$R_T = \frac{\rho k^2}{\mu\varepsilon} \qquad (9)$$

The measurements in the plane of the impeller as used as model boundary conditions were ensemble-average velocities and thus the periodic component was included in the measurements. All calculations were carried out on three dimensional grids in cylindrical coordinates using the third order accurate QUICK interpolation scheme. Further details of the modeling conditions are given by Sturesson and Rasmuson [6].

RESULTS AND DISCUSSIONS

Experimental LDA-measurements of the mean and fluctuating velocity components were performed for three impeller rotational speeds (N= 3, 6 and 9 rev/s) and compared with numerical results at selected locations in the near wall and bottom regions. It was found that despite the presence of the solid boundary the mean and fluctuating velocities scale quite well with impeller tip speed for the three impeller speeds, although the dimensionless velocity values for N= 3 rev/s had a tendency to be less than for N= 6 and 9 rev/s. Measurements near the bottom and walls were conducted upto a distance of about 5 mm from the solid surfaces. Further from the solid surface the experimentally determined flow properties were found to be in agreement with the bulk flow measurements for the same system as determined by Sturesson and Rasmuson [6]. Numerical predictions were performed for N= 9 rev/s and shown as lines in the Figures. Dotted lines corresponds to using wall functions and solid lines to the low Reynolds number model. The axial fluctuating velocity components are not reported for the measurements near the vertical wall because of the extraneous light reflected from the wall of the tank as discussed above.

Near Bottom Region

The radial and tangential velocities were measured at a radial distance of 70 mm from the vessel centerline. Measurements were conducted from 50 µm to 5 mm above the bottom surface.

Mean velocities: The experimentally determined radial and tangential mean velocity profiles for three impeller speeds are shown in Figures 4 and 5. The lines correspond to the numerical predictions for N=9. The declining velocity profile close to the solid surface, typical for a turbulent boundary layer, can clearly be distinguished.

The dimensionless radial mean velocity is approximately three times the value of the dimensionless tangential mean velocity. Numerical predictions are qualitatively in agreement with the experimental data. However, the obtained numerical values are translated to lower mean values possibly due to an incorrect outer value of the model further from the solid surface. The difference between the two different wall models are relatively marginal. It is worth noting that the grid distribution was close to the solid surface when using the low Reynolds number model which makes a detailed comparison between the two different near wall models difficult.

The resulting mean velocity parallel to the bottom surface is shown in Figure 6 for all three rotational speeds. For N= 3 rev/s the velocity profile exhibit a constant slope to a distance of about 0.1 mm above the bottom. The measuring position closest to the wall (50 µm) can for the lowest impeller speed thus be considered to be located in the viscous sublayer. The shear stress can then be obtained by a linear extrapolation to the solid surface. The shear-stress obtained in the same manner for N= 6 and 9 rev/s have to be viewed with caution since the thickness of the viscous sublayer is decreased when increasing the impeller speed [1].

The experimentally obtained values of the shear-stress were 0.7 N/m^2, 2.5 N/m^2 and 4.6 N/m^2 for N= 3, 6 and 9 rev/s, respectively. These results indicate a linear increase of the wall shear stress at the bottom with respect to the impeller speed. The dimensionless frictional velocity and distance from the surface can then be calculated according to: $u^* = \sqrt{\tau_w/\rho}$ and $y^+ = u^* y/\nu$. The dimensionless mean velocity ($u^+ = U_{par}/u^*$) plotted against the dimensionless distance from the bottom for N= 3, 6 and 9 rev/s are shown in Figures 7a-c.

The linear slope expected in the viscous sublayer fits the measured positions closest to the wall. However, the outer part is not consistent with the logarithmic feature typical for a turbulent boundary layer over a flat plate. A logarithmic line fitted to the obtained experimental data would yield a slope inconsistent with what can be expected from boundary layers over flat plates. This fact points out the added complexity of the rotational flow in a stirred tank compared to the idealized flow over a flat plate.

Fluctuating velocities: Plots of the dimensionless root mean square fluctuating velocities (RMS) parallel to the bottom surface are shown in Figures 8 and 9.

It can be seen that the magnitude of the fluctuating tangential component is greater than the radial counterpart, despite the fact that the mean radial velocity was greater than the tangential mean velocity. Both the radial and tangential fluctuating velocity components exhibit a peak value close to the solid surface. In Figures 10a-c the tangential and radial components are plotted separately for each impeller rotational speed using the dimensionless distance (y^+) from the bottom.

A peak in the fluctuating velocities is clearly recognized for all three impeller speeds. According to experimental data for flow over flat plates (e.g. Hinze [9]) the peak is expected in the bufferlayer. The peak is found at $y^+ = 25$ for N= 3 rev/s. However, for N= 6 and 9 rev/s the peak is found at a larger value of y^+ of about 40. This can be due to the fact that the shear-stress was not properly determined for these impeller speeds as noted above. Moreover, the fluctuating values coincide further from the wall indicating an isotropic behavior away from the solid surface.

Near Wall Region

The axial velocity profiles close to the vessel wall were measured at four heights above the vessel bottom. The axial velocity component was measured using the length of the measuring control volume normal to the front wall which decreases the spatial resolution (with respect to the distance from the solid surface) compared to when using the width of the measuring control-volume. The fluctuating axial velocity component is not reported due to the light reflected from the vertical wall. The experimental results of the mean axial velocity together with the numerical calculations are shown in Figures 11-14.

At the height of 5 mm above the bottom the circulation loop at the corner of the vessel is clearly seen and the flow starts to develope along the vertical wall. Further above the base the fluid is forced upwards. At the height of 20 mm above the bottom the velocity profile is rather flat. At 30 mm above the bottom the mean axial velocity declines at about 3 mm from the vessel wall.

Measurements of the radial and tangential components were conducted with the beams parallel to the bottom surface. In this case the width of the control-volume determined the closest measuring position. The axial velocity component can be measured close to the wall using the width of the control-volume in the plane perpendicular to the incoming laser beams. However, this approach involves measurements close to a strong vessel curvature and would require index matching of the fluid in the square tank. Measurements as such have not been attempted in this study.

CONCLUSIONS

It can be concluded that it was possible to measure the turbulent distributions close to the solid surface of a stirred tank using LDA-technique. It was found that despite the presence of the solid boundary the mean and fluctuating velocities scale with impeller tip speed for the three impeller speeds studied, although the dimensionless velocity values for N= 3 rev/s had a tendency to be less than for N= 6 and 9 rev/s. The measurements reveal typical characteristics for a turbulent boundary layer over a the solid surface. The linear mean velocity profile in the viscous sublayer together with the peak of the fluctuating velocities close to the buffer layer have been shown. The development of the mean axial flow field along the vertical wall was characterized. Numerical predictions show qualitative agreement although the numerical values underpredicts the mean velocity profiles. This fact is possibly due to an incorrect value of the model further from the wall.

NOMENCLATURE

B - width of baffles (m)
C - impeller clearance (m)
D - impeller diameter (m)
H - height of clear liquid (m)
k - turbulent kinetic energy per unit mass (m^2/s^2)
N - impeller speed (rev/s)
N_{Re} - mixing Reynolds number (ND^2/ν) (-)
r - radial coordinate (m)
RMS - dimensionless root mean square fluctuating velocity (-)
T - vessel diameter (m)
u - fluctuating velocity component (m/s)
u^+ - dimensionless velocity (-)
U - fluid velocity (m/s)
$\overline{U}$ - mean velocity (m/s)
U - dimensionless mean axial velocity (-)
U_{par} - mean velocity parallel to the bottom surface (m/s)
V - dimensionless mean radial velocity (-)
VRMS - dimensionless radial root mean square fluctuating velocity (-)
W - dimensionless mean tangential velocity (-)
WRMS - dimensionless tangential root mean square fluctuating velocity (-)
y^+ - dimensionless distance from solid surface (-)
z - axial coordinate (m)

Greek symbols

α - angle between incident beams (°)
δ_{ij} - Kronecker delta
ε - turbulent dissipation (m^2/s^3)
θ - azimuthal direction (°)
λ - wavelength of light (nm)
μ - molecular viscosity (Pas)
μ_t - turbulent viscosity (Pas)
ν - kinematic viscosity (m^2/s)
ρ - density (kg/m^3)
τ_w - wall shear stress (N/m^2)

LITERATURE CITED

1. Sturesson, C.; Rasmuson, A., Complete Suspension of Solid Particles in Stirred Vessels, *Proc. 8th Europ. Conf. on Mixing*, Cambridge, UK, 21-23 Sept., IChem Symp. Series **136**, 357, (1994a).

2. Fort, I.; Vlcek, J.; Cink, M.; Hruby, M., Velocity Field at the Wall of Vessel with Axial Mixer and Radial Baffles, *Coll. Czech. Chem. Comm.*, **36**, 1546, (1971).

3. Placek, J.; Tavlarides, L.L.; Smith, G.W.; Fort, I., Turbulent Flow in Stirred Tanks Part II: A Two-Scale Model of Turbulence, *AIChE J.*, **32**, 1771, (1986).

4. Rosén, C.; Trägårdh, C., $\overline{u'v'}$ Reynolds Stress in the Viscous Sublayer over a Wide Range of *Re* Numbers, *AIChE J.*, **40**, 29, (1994).

5. Johansson, T.G.; *An Experimental Study of the Structure of a Flat Plate Turbulent Boundary Layer, Using Laser-Doppler Velocimetry*, PhD thesis, Chalmers University of Technology, Gothenburg, Sweden, (1988).

6. Sturesson, C.; Rasmuson, A., An Experimental and Computational Investigation of the Mean and Turbulent Flow Field Produced by a Pitched Blade Impeller in a Fully Baffled Vessel, submitted for publication to *Chem. Eng. Sci.*, (1994b).

7. Launder, B.E.; Spalding, D.B., The Numerical Computation of Turbulent Flows, *Comp. Meths. Appl. Mech. Eng.*, **3**, 269, (1974).

8. Sturesson, C.; Rasmuson, A.; Fort, I, Numerical Simulation of Turbulent Flow of Agitated Liquid with Pitched Blade Impeller, in press *Chem. Eng. Comm.*, (1995).

9. Hinze, J.O.; *Turbulence*, McGraw-Hill, New York, (1975).

TABLE 1. Characteristics of the measuring control volume.

	$\lambda = 514.5$ nm	$\lambda = 488$ nm
length, in air (mm)	0.546	0.498
width (mm)	0.062	0.058
height (mm)	0.062	0.059
fringe spacing (µm)	2.281	2.082

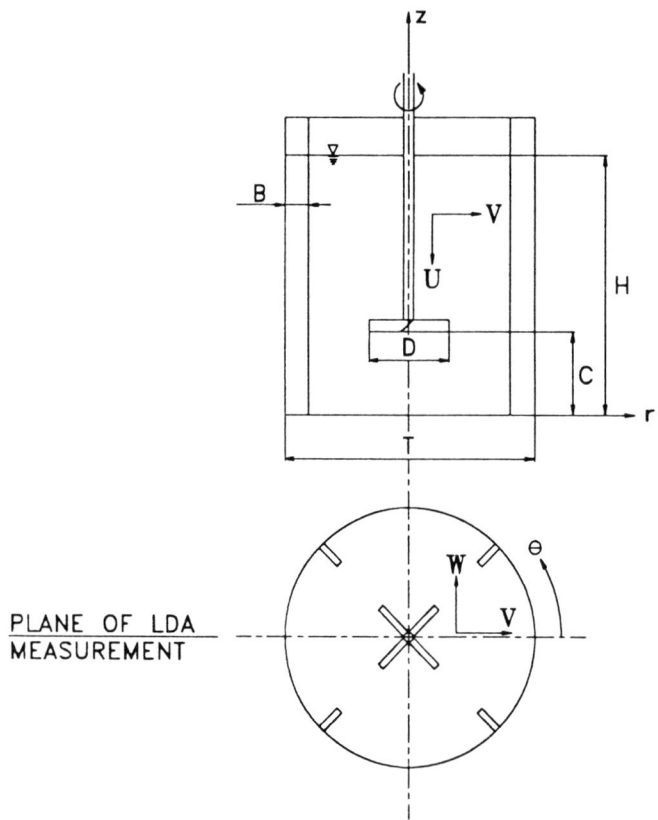

Figure 1. Geometry of the system.

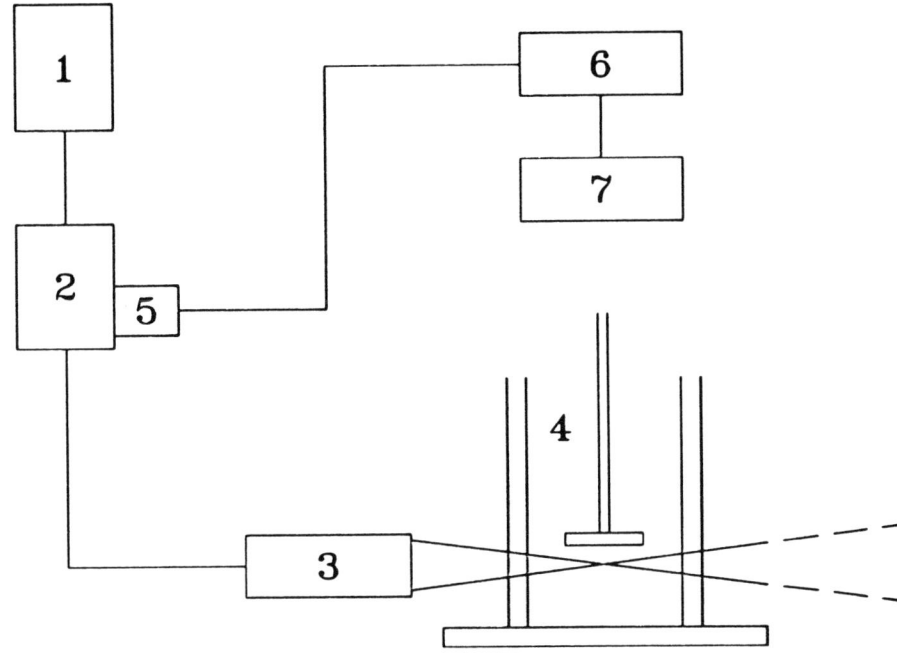

Figure 2. The LDA-setup: 1-Argon-ion laser, 2-Transmitting optics, 3-Fiber probe including receiving optics, 4-Test section, 5-Photomultipliers, 6-Signal processors, 7-Computer.

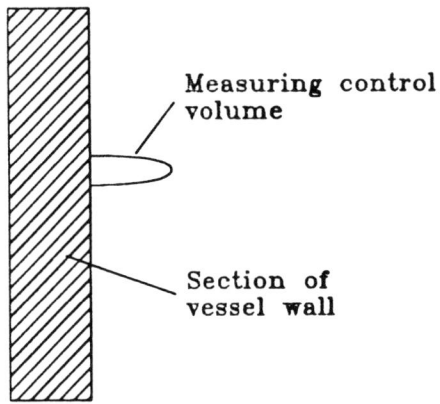

Figure 3. Measurement position close to wall.

110   Industrial Mixing Fundamentals with Applications

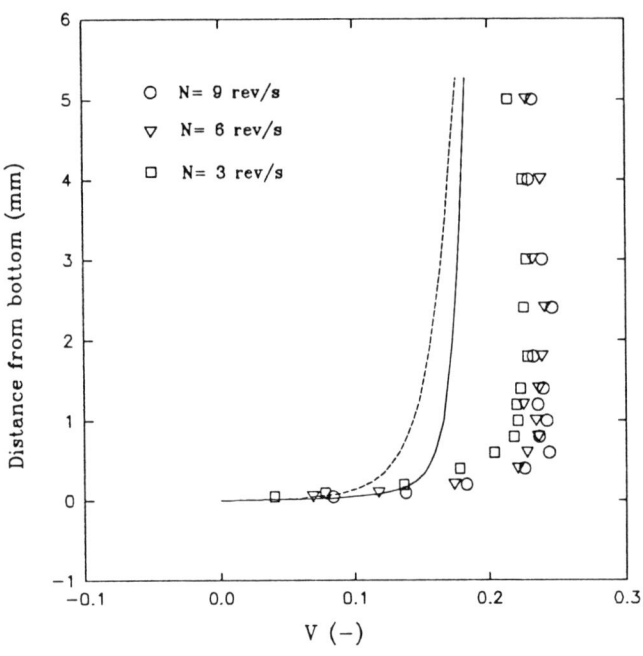

Figure 4. Mean radial velocity profile at r= 70 mm.

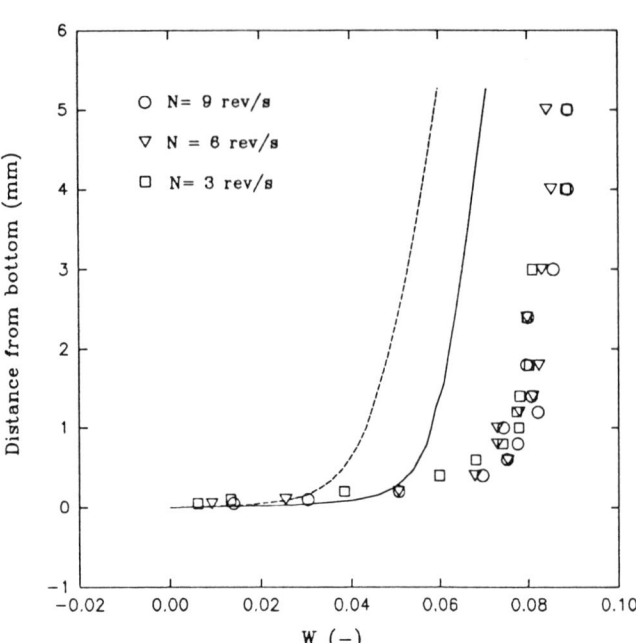

Figure 5. Mean tangential velocity profile at r= 70 mm.

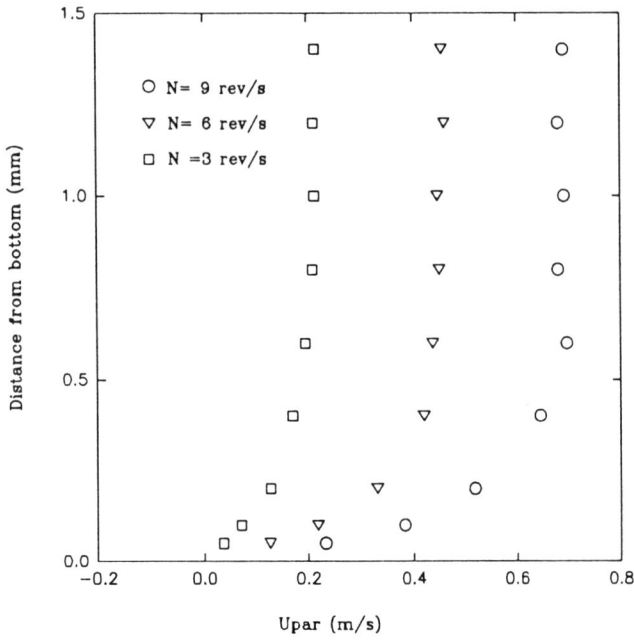

Figure 6. Mean velocities parallel to the bottom at r= 70 mm.

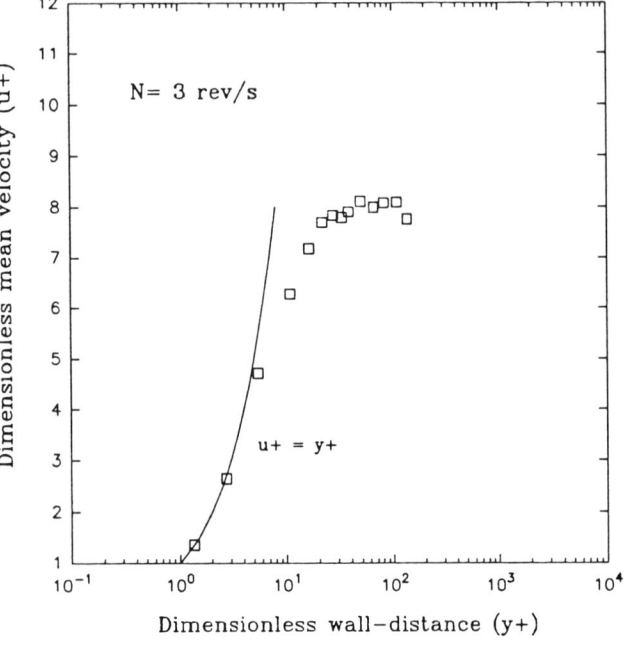

Figure 7a. Dimensionless velocity (N= 3 rev/s).

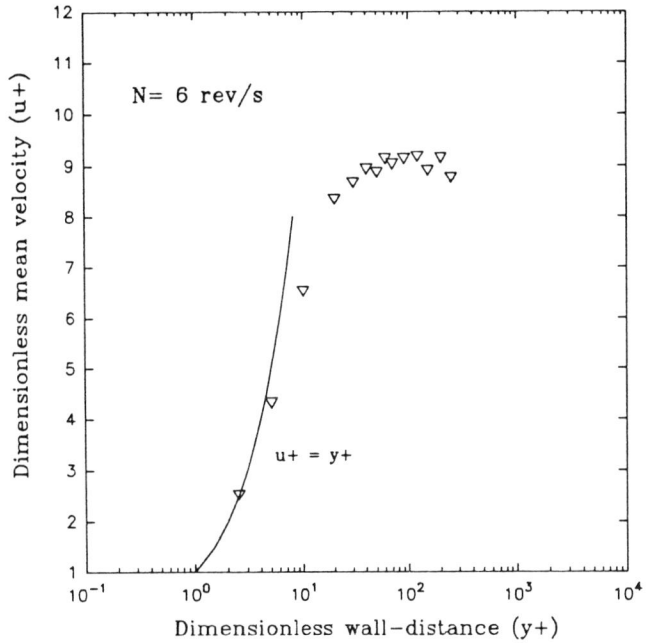

Figure 7b. Dimensionless velocity (N= 6 rev/s).

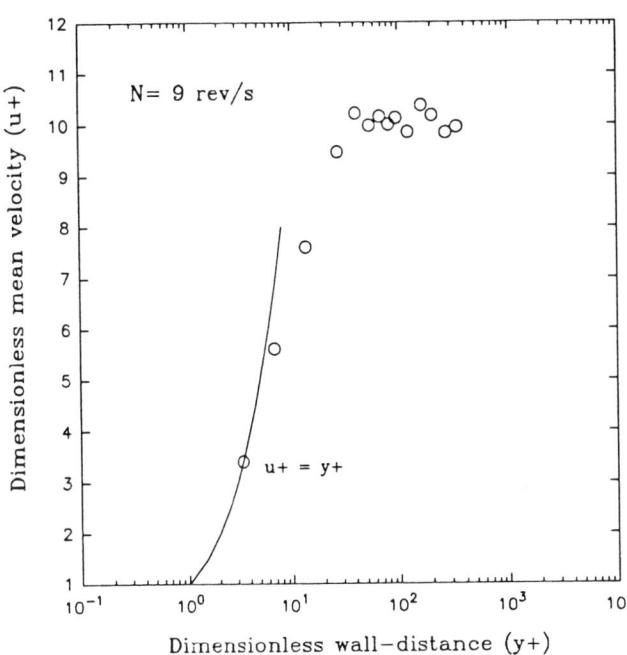

Figure 7c. Dimensionless velocity (N= 9 rev/s).

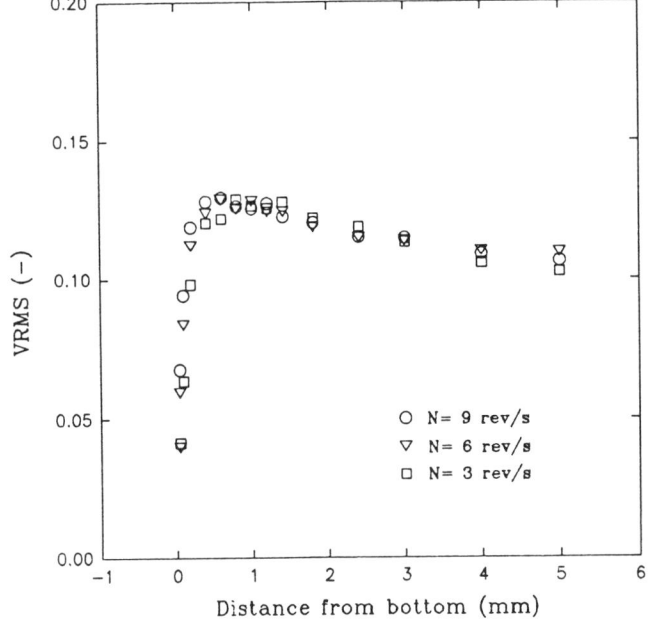

Figure 8. Radial fluctuating velocities.

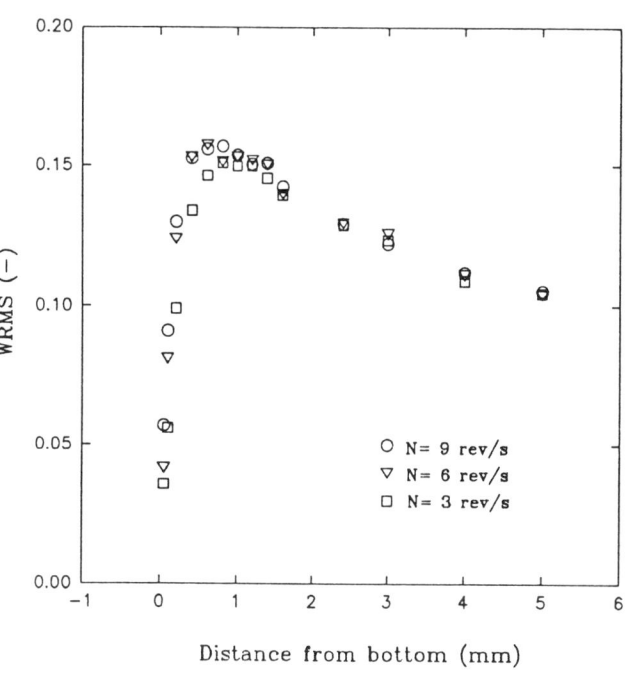

Figure 9. Tangential fluctuating velocities.

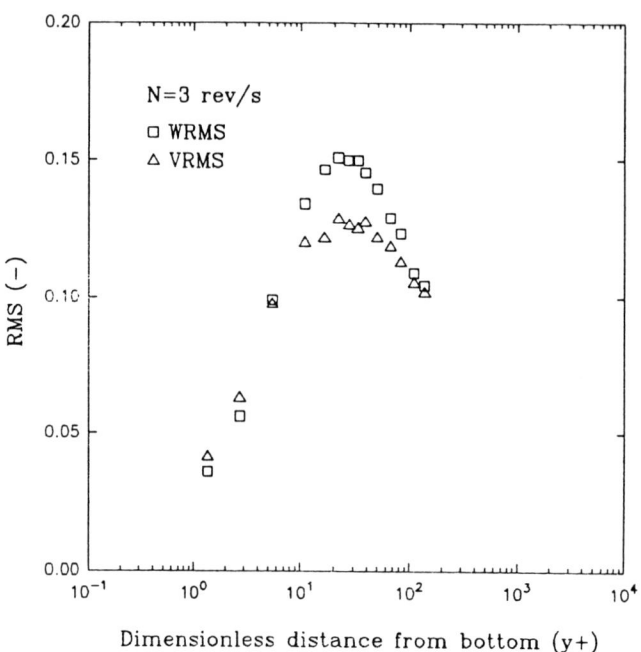

Figure 10a. Radial and tangential fluctuating velocities (N= 3).

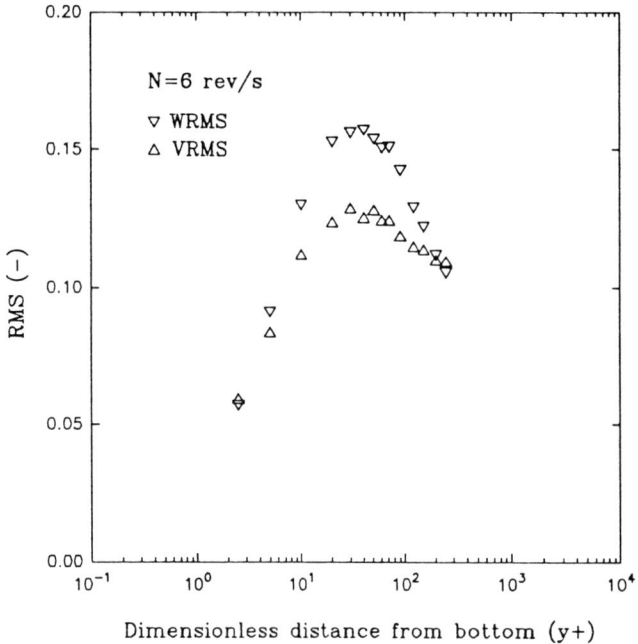

Figure 10b. Radial and tangential fluctuating velocities (N= 6).

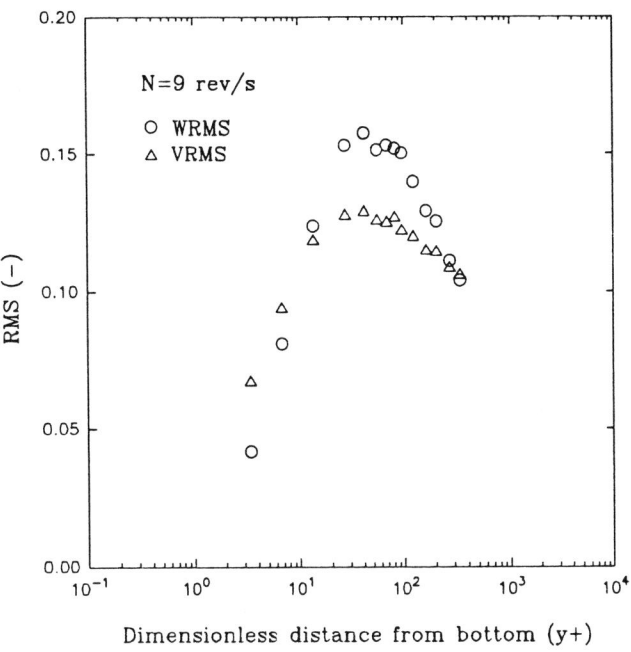

Figure 10c. Radial and tangential fluctuating velocities (N= 9).

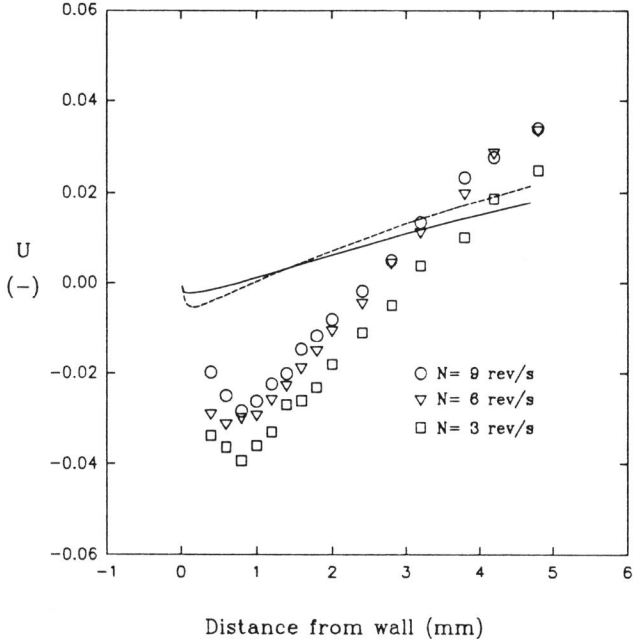

Figure 11. Mean axial velocity at 5 mm above the bottom.

114   Industrial Mixing Fundamentals with Applications   AIChE SYMPOSIUM SERIES

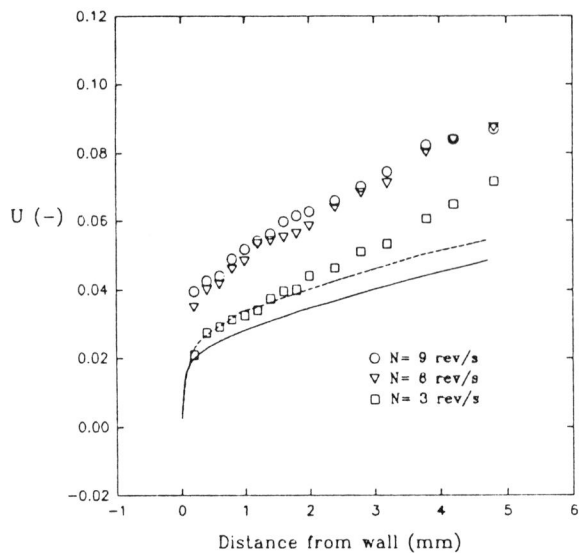

Figure 12. Mean axial velocity at 10 mm above the bottom.

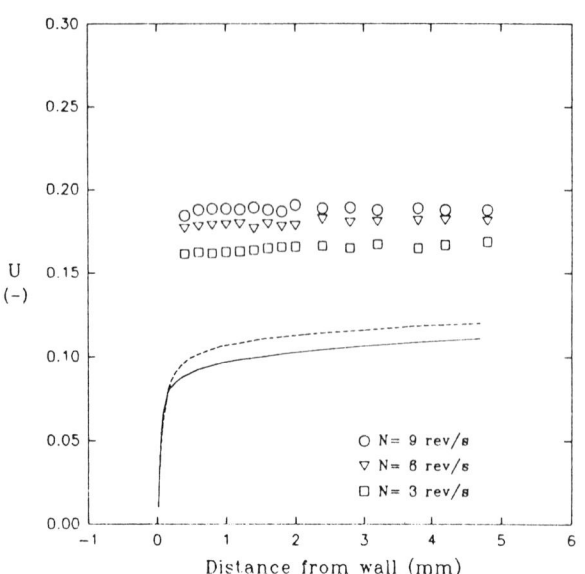

Figure 13. Mean axial velocity at 20 mm above the bottom.

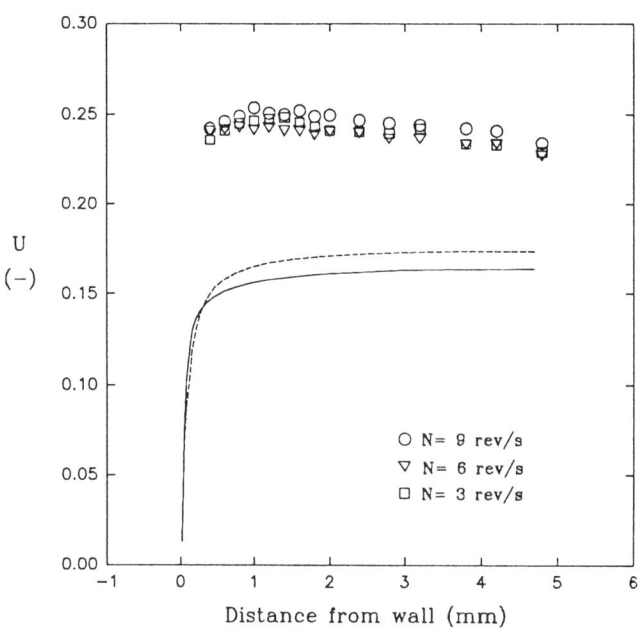

Figure 14. Mean axial velocity at 30 mm above the bottom.

On the Effect of Wall and Bottom Clearance on Mixing of Viscoelastic Fluids

Jianya Cheng, Pierre J. Carreau
Centre de recherche appliquée sur les polymères, CRASP, Department of Chemical Engineering
Ecole Polytechnique, C.P. 6079, Succursale Centre-Ville, Montreal, Quebec, Canada H3C 3A7

Raj P. Chhabra
University of New South Wales, School of Chemical Engineering and Industrial Chemistry
Sydney NSW, 2052, Australia

The clearance between the vessel wall and the impeller blade is an important design parameter which may considerably affect the mixing performance of an agitated system. The effect of wall clearance on the power requirement and on the effective deformation rate has been investigated for viscoelastic fluids mixed by helical ribbon agitators. The power input and effective deformation rate are shown to increase considerably with decreasing wall clearance. The effective deformation rate, at large values of wall clearance, is apparently independent of wall clearance. A model based on the equivalent Couette flow analogy is proposed to predict the effective deformation rate for inelastic shear-thinning fluids in the laminar flow regime. It includes the effects of both side wall clearance and bottom clearance. The model predicts well the influence of wall clearance on the effective deformation rate constant, k_s. The bottom clearance is shown to have little effect on the power consumption and on the effective deformation rate.

Mixing of high viscosity Newtonian and viscoelastic fluids is encountered in many industrially important processes. Typical examples include blending of viscous petroleum fractions, polymeric melts with additives in polymer processing applications, fermentation broths in biotechnological processes. While the mixing of low viscosity systems is accomplished by using high speed agitators under turbulent regime in the vessel, most high viscosity Newtonian and non-Newtonian systems are mixed under laminar conditions. Under these conditions, the momentum imparted by the rotating impeller to the fluid is not transmitted very far and thus, a key to the efficient mixing is to employ a close clearance impeller with large bulk flow or turn over. One such device which has proved highly successful and has thus gained wide acceptance for the mixing of high viscosity systems is the so called helical ribbon (HR) impeller [1,2,3].

It is readily agreed that the power consumption for achieving a desired level of mixing represents one of the most important design parameters in such applications. For non-Newtonian fluids, considerable attention has also been given to the estimation and/or prediction of the effective shear rate prevailing in the vessel. Both these aspects, namely, power consumption and effective shear rate are not only inter-related but are also complementary to each other. For instance, the knowledge of the effective deformation rate facilitates the calculation of the effective viscosity in such a manner that the power curve developed for Newtonian fluids can also be used to estimate the power input to mix a non-Newtonian fluid for the same geometrical configuration [4]. It is generally recognized that both power consumption and effective deformation rate are strongly dependent on a large number of variables including the rheological characteristics (shear rate dependent viscosity, viscoelasticity, etc.), operation conditions (laminar or turbulent regime) as well as on the various geometric ratios of the impellers/vessel combination [3,5]. While much work has been carried out regarding the role of rheological properties on power consumption of helical ribbon impellers [3,6], little is known about the influence of the geometrical parameters on the performance of helical ribbon in mixing high viscosity systems. One of the most important geometric parameters exerting strong influence on power consumption is the side wall clearance, i.e., the gap between the vessel wall and the rotating impeller. This has been shown to be a significant variable even for the mixing of Newtonian systems [7]. Likewise, very little is know about the effect of bottom clearance on power consumption. The present work aims to fill this gap in the existing literature. It is, however, instructive to begin with by providing a terse account of the pertinent previous studies in this area.

PREVIOUS WORK

The available large body of information on the influence of geometric ratios of impeller-vessel combinations for low viscosity systems has been critically reviewed amongst others by Nagata [8] and Oldshue [9]. Here, only those relating to helical ribbon impellers used for high viscosity systems are mentioned briefly. Hall and Godfrey [10] measured the power input data for the mixing of non-Newtonian media by helical ribbon impellers and found it to be strongly influenced by the side wall clearance (c/d). Similar conclusions have been since presented by Nagata [8]. Kappel [7] provided a succinct summary of the available studies pertaining to the influence of various geometric parameters on power consumption of helical ribbons. Owing to intensive shearing of fluid in the thin gap, the power input rises with decreasing side wall clearance even for Newtonian fluids. The effect is, however, likely to be less dramatic for shear-thinning fluids due to the concomitant decrease in effective viscosity. Kappel [7], however, concluded that an omission of side clearance allowance in the estimation of power can entail errors up to 100%. Takahashi et al., [11] showed that the effective shear rate showed a strong dependence on the side wall clearance for shear-thinning materials. Subsequently, Ayazi Shamlou and Edwards [12] also reached similar conclusions regarding the effect of side wall clearance on both power consumption and the effective deformation rate. The effect of bottom clearance has received even less attention. Only recently, Carreau et al. [1] demonstrated the effect to be rather small for Newtonian fluids.

It is thus fair to re-iterate that little is known about the side wall and bottom clearance effects on the effective deformation rate and power consumption for mixing of high viscosity Newtonian and non-Newtonian fluids by helical ribbon impellers. In this work, based on a simple model, a theoretical framework is presented which elucidates the influence of the side and bottom clearances on the power consumption and effective shear rate involved in the mixing of power-law fluids by helical ribbon impellers. The theoretical predictions made here in have been substantiated by experimental data for Newtonian and non-Newtonian fluids.

THEORETICAL CONSIDERATIONS

A schematic representation of a helical ribbon agitator together with the definitions of various geometrical aspects is shown in Figure 1. Following the postulates of Bourne and Butler [13] and Ulbrecht and Carreau [6], the fluid motion induced by the rotating helical ribbon impeller is approximated by an equivalent flow produced in a coaxial cylinder configuration with the inner cylinder of an equivalent diameter rotating. It is assumed that the total energy dissipation (power consumption) is due to the shearing of the fluid between the annular gap (side clearance) and that in between the base of the impeller and the bottom of the wall and that these two components are additive. We begin with the estimation of power or torque due to the shearing in the annular gap.

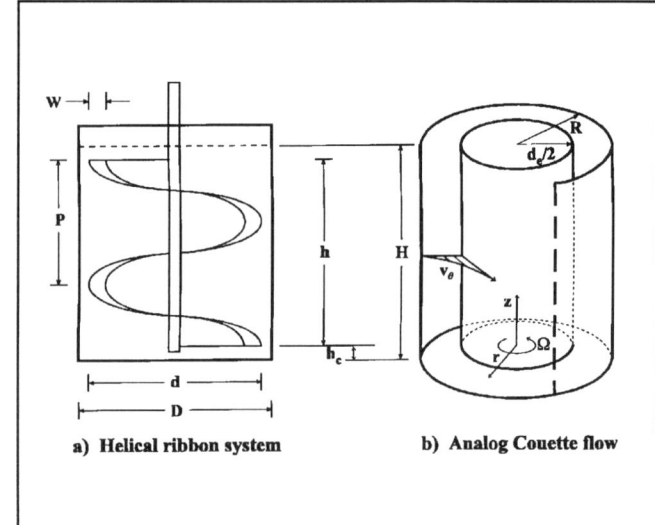

Figure 1   Sketch of mixing setup. a) Helical ribbon agitator b) Equivalent Couette geometry

Power Consumption

For this flow configuration, it is reasonable to postulate that the fluid is subjected to unidirectional shearing motion and the kinematics of the flow can be described by:

$$v_r = v_z = 0 \qquad (1)$$

$$v_\theta = v_\theta(r) \qquad (2)$$

$$\sigma_{rz} = \sigma_{\theta z} = 0 \qquad (3)$$

$$\sigma_{r\theta} = f(r) \qquad (4)$$

Under these conditions and noting that there is no pressure gradients, the θ-component of the equation of motion simplifies to:

$$\frac{\partial}{\partial r}(r^2 \sigma_{r\theta}) = 0 \qquad (5)$$

For the power-law fluids, the shear stress is given by:

$$\sigma_{r\theta} = -m(\dot{\gamma}_{r\theta})^n = -m\left[r\frac{\partial}{\partial r}\left(\frac{v_\theta}{r}\right)\right]^n \qquad (6)$$

Combining equations (5) and (6), integrating and using the boundary conditions: $v_\theta = 0$ at $r = D/2$ and $v_\theta = \Omega\, d_e/2$, one obtains:

$$\frac{v_\theta}{r} = \frac{\Omega}{\left\{\left(\frac{D}{d_e}\right)^{\frac{2}{n}} - 1\right\}} \left\{\left(\frac{R}{r}\right)^{\frac{2}{n}} - 1\right\} \qquad (7)$$

and the corresponding shear stress can be obtained by substituting (7) in (6) as:

$$\sigma_{r\theta} = -m\left[\frac{2\Omega}{n\left[\left(\frac{D}{d_e}\right)^{\frac{2}{n}} - 1\right]}\right]^n \left(\frac{R}{r}\right)^2 \qquad (8)$$

The torque exerted on the vertical wall at the vessel is obtained:

$$\Gamma_{wall} = (2\pi R H)(-\sigma_{r\theta})|_{r=R}\, R \qquad (9)$$

as

$$\Gamma_{wall} = \frac{\pi}{2} D^2 H m \left[\frac{2\Omega}{n\left[\left(\frac{D}{d_e}\right)^{\frac{2}{n}} - 1\right]}\right]^n \qquad (10)$$

Flow Between the Base of Impeller and Bottom of Vessel

In this case, following the consideration advanced by Bird et al. [14] for a similar flow situation, we neglect the r-and z-components of the flow field, i.e.,

$$v_r = v_z = 0 \qquad (11)$$

Under these conditions the equation of continuity suggests:

$$v_\theta = r\, f(z) \qquad (12)$$

Likewise, one can argue that $\sigma_{r\theta} \ll \sigma_{z\theta}$ and therefore the θ-component of the equation of motion reduces to:

$$\frac{\partial \sigma_{z\theta}}{\partial z} = 0 \qquad (13)$$

For this flow configuration, the power-law model may be expressed as:

$$\sigma_{z\theta} = -m(\dot{\gamma}_{z\theta})^n = -m\left(\frac{\partial v_\theta}{\partial z}\right)^n = c_1 \qquad (14)$$

Integrating equation (14) using the boundary conditions, at $z=0$, $v_\theta = 0$ and at $z = h_c$, $v_\theta = \Omega r$ one obtains:

$$V_\theta = \frac{\Omega r z}{h_c} \qquad (15)$$

The shear stress in then given by

$$\sigma_{z\theta} = -m\left(\frac{\Omega}{h_c}\right)^n r^n \qquad (16)$$

The torque due to the shearing in this region is obtained as:

$$\Gamma_{bottom} = \int_0^{d_e/2} -r\,\sigma_{z\theta}(r)\, 2\pi r\, dr$$

$$= \frac{2\pi m}{(n+3)}\left(\frac{d_e}{2}\right)^{n+3}\left(\frac{\Omega}{h_c}\right)^n \qquad (17)$$

The total torque on the inner rotating cylinder (or ribbon impeller) is simply the sum of two contributions given by equations (10) and (17), i.e.,

$$\Gamma_{total} = \frac{\pi}{2} D^2 H m \left[\frac{2\Omega}{n\left\{\left(\frac{D}{d_e}\right)^{2/n} - 1\right\}}\right]^n$$

$$+ \frac{2\pi m}{n+3}\left(\frac{\Omega}{h_c}\right)^n\left(\frac{d_e}{2}\right)^{n+3} \qquad (18)$$

The power P input to the fluid is given by
$$P = \Omega \, \Gamma_{total} \quad (19)$$
$$= 2\pi N \, \Gamma_{total}$$
Substitution of equation (18) into (19) yields:
$$P = m\pi^{n+2} N^{n+1} \left\{ D^2 H \left[\frac{4}{n\left[(D/d_e)^{2/n} - 1\right]} \right]^n \right.$$
$$\left. + \frac{D^3}{2(n+3)} \left(\frac{H}{h_c}\right)^n \left(\frac{D}{H}\right)^n \left(\frac{d_e}{D}\right)^{n+3} \right\} \quad (20)$$

Hence, equation (20) provides the theoretical framework for the elucidation of the effects of side and bottom clearances on the power input to the fluid.

Effective Shear Rate

Following the approach introduced by Metzner and Otto [4], the effective shear rate in the mixing vessel is given by:
$$\dot{\gamma}_{eff} = k_s \, N \quad (21)$$
which in turn facilitates the calculation of the effective viscosity for a power law fluid as:
$$\eta_{eff} = m \left(\dot{\gamma}_{eff}\right)^{n-1} \quad (22)$$
One can now define a generalised Reynolds number in usual manner as:
$$Re_g = \frac{\rho \, Nd^2}{\eta_{eff}} \quad (23)$$
This approach has the virtue that one can use the power curve obtained with Newtonian fluids for a geometrically similar system. For a given geometrical configuration, the power number, Np, varies inversely proportional to the Reynolds number, Re_g, in the laminar region, that is:
$$Np \, Re_g = K_p \quad (24)$$
where K_p is a constant characteristic of the impeller/vessel combination. After substitution for $Np = P/\rho N^3 d^5$, equation (24) can be rearranged to yield the following expression for η_{eff}:
$$\eta_{eff} = \frac{P}{Kp \, d^3 \, N^2} \quad (25)$$

From equations (22) and (25), it can be inferred that
$$\dot{\gamma}_{eff} = \left(\frac{P}{m \, Kp \, d^3 \, N^2}\right)^{\frac{1}{n-1}} \quad (26)$$
and finally, combining equations (20) and (26), one obtains:
$$\frac{\dot{\gamma}_{eff}}{N} \equiv k_s = \left\{\frac{\pi^{n+2}}{K_p \, d^3}\right\}^{\frac{1}{n-1}} \left[D^2 H \left\{\frac{4}{n\left[(D/de)^{2/n} - 1\right]}\right\}^n \right.$$
$$\left. + \frac{D^3}{2(n+3)} \left(\frac{H}{h_c}\right)^n \left(\frac{D}{H}\right)^n \left(\frac{d_e}{D}\right)^{n+3} \right]^{\frac{1}{n-1}} \quad (27)$$

Thus, equation (27) can be used to delineate the complex dependence of the effective shear rate on the side clearance (d_e/D) and the bottom clearance (H/h_c) and the possible interplay between the fluid rheology and system geometry.

In the absence of bottom effects, equation (27) reduces to the result derived by Carreau et al., [1]. This study also suggested that the equivalent diameter, d_e, varies somewhat with the flow behaviour index (n), even for the same geometrical configuration. Based on their results obtained with six different helical ribbons impellers, it is reasonable to assume that the equivalent diameter, d_e, for a given power-law fluid bears almost a constant ratio (within $\pm 10\%$) with that for Newtonian fluids irrespective of the mixer geometry, i.e.,
$$\frac{d_e \, (n \neq 1)}{d_e \, (n = 1)} = f(n) \quad (28)$$
where $f(n)$ can be estimated from the results of Carreau et al., [1]. Thus the equivalent diameter for a shear-thinning fluid can be estimated via equation (28), calculating d_e from power data for a Newtonian fluid via equation (20) with $n = 1$. We now turn our attention to the experimental part of this work.

EXPERIMENTAL

Since the detailed description of the experimental setup, procedure and materials are available elsewhere [1,2,15], only the salient features are recapitulated here. The same test fluids whose rheological properties have been presented by Carreau et al. [1]

were employed in this study. The helical ribbon impellers were of the same diameter, d equal to 0.263 m, and three vessels were used ($D = 0.279$, 0.292 and $0.381 m$) to investigate the effect of side wall clearance which varied from 0.029 to 0.155 m. Likewise, the effect of bottom clearance on power consumption was investigated for three values of h_c as $10 mm$, $50 mm$ and $85 mm$.

RESULTS AND DISCUSSION

Effect of Side Wall Clearance

As outlined in the theoretical section above, initially the value of Kp was deduced from the power measurement with a Newtonian corn syrup as a function of Reynolds number for each side clearance. Figure 2 shows the typical power number - Reynolds number behavior for three values of the side clearance, for a constant value of $h_c = 100 mm$.

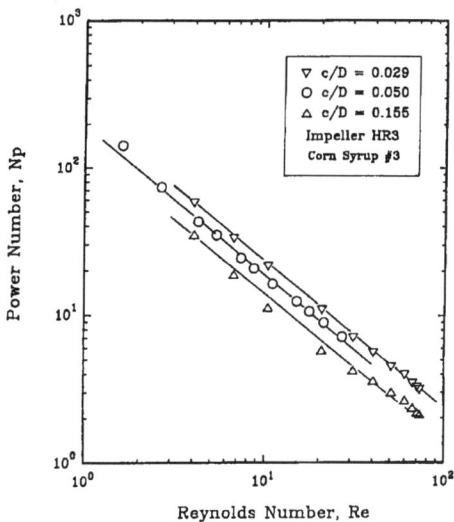

Figure 2   Power number as a function of Reynolds number for a Newtonian corn syrup ($\mu = 2.14 Pa.s$, $\rho = 1280 kg/m^3$) for three different wall clearance values.

The increase in power consumption with the decreasing value of (c/D) is simply attributable to the increasing level of effective shearing in the annular gap. Qualitatively similar results have been reported by Hall and Godfrey [10]. The resulting values of K_p ($= N_p Re$) are summarized in Table 1 for the range of conditions covered in this study. For a fixed geometry of the impeller, the value of K_p is seen to increase with the decreasing side wall clearance; the extent of increase is, however, strongly dependent on the width and pitch of impellers, as shown for two impellers in Figure 3 (only one result has been obtained for impeller HR1, not reported in the figure).

Table 1. Influence of side wall clearance on K_p

No	c/D	Impeller	K_p exp.	K_p Chavan and Ulbrecht	K_p Yap et al.	K_p Shamlou and Edwards
1	0.029	HR1	--	239.1		283.1
2		HR2	217.2	227.4		234.2
3		HR3	227.6	266.3		413.8
4	0.050	HR1	164.1	180.9	136.2	226.7
5		HR2	132.2	176.7	103.0	229.5
6		HR3	192.2	206.8	129.8	253.9
7	0.155	HR1	--	92.7		161.0
8		HR2	108.0	96.4		169.7
9		HR3	132.1	112.1		179.2

Table 1 also reports the values of K_p predicted by three different models proposed in the literature [12,16,17]. Note the good correspondence between our data and the predictions of the model of Chavan and Ulbrecht [16]. The predictions using the model of Ayazi Shamlou and Edwards [12] are consistently higher, in some cases almost twice the experimental values.

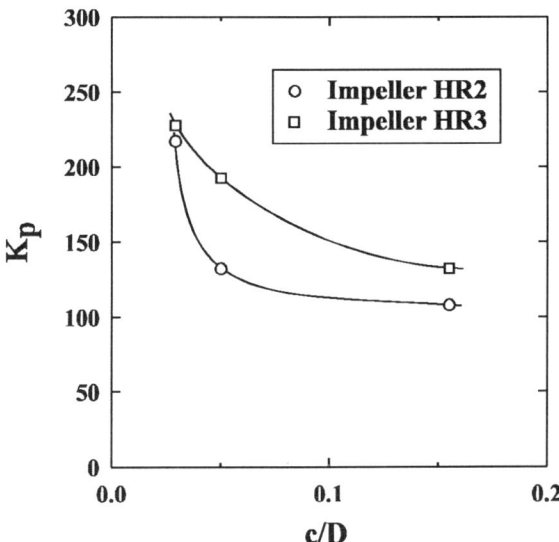

Figure 3   Effect of side wall clearance on the power constant Kp, for a Newtonian corn syrup (bottom clearance, $h_c = 10 mm$).

Typical results illustrating the effect of side clearance on the effective deformation rate, in terms of k_s, is shown in Figure 4 for an inelastic carboxymethyl cellulose solution (The Cross model parameters are: $n = 0.3$, $\eta_0 = 483 Pa \cdot s$, $t_1 = 7.83 s$, see Carreau et al., [1]). Results for two helical ribbons are shown. As expected, the effective deformation rate increases with decreasing side clearance. Included in this figure are the predictions of equation (27) and the agreement between the experiments and predictions is seen to be about as satisfactory as can be expected in this kind of work. Also shown are the results of Ayazi Shamlou and Edwards [12] and Takahashi et al., [11]. Based on the literature as well as their own data, Ayazi Shamlou and Edwards [11] presented the following correlation for k_s (in the range $0.068 \leq (c/d) \leq 0.164$):

$$k_s = 34 - 144 \left(\frac{c}{d} \right) \quad (29)$$

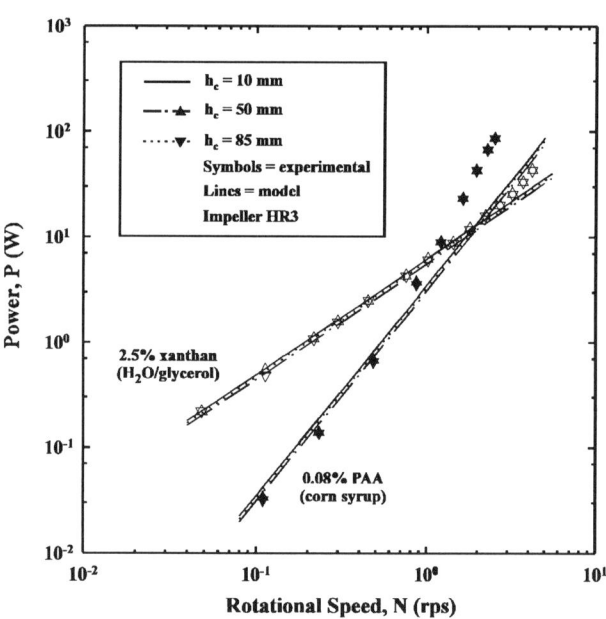

Figure 4   Effect of side wide clearance on k_s for a 3% CMC solution (bottom clearance, $h_c = 10 mm$).

The predictions of equation (28) are in fair agreement with the present results at least for low clearance values. This correlation cannot account for effect of impeller geometry which is clearly shown in the figure. Based on wide ranges of rheological properties ($0.44 \leq n \leq 0.77$) and side clearance ($0.023 \leq (c/d) \leq 0.097$), Takahashi et al., [11] proposed a correlation which takes also into account the pitch and the blade width:

$$k_s = 11.4 \left(\frac{c}{D} \right)^{-0.411} \left(\frac{p}{D} \right)^{-0.361} \left(\frac{w}{D} \right)^{0.164} \quad (30)$$

Clearly, their predictions, shown by the dashed lines in Figure 4, are too high by about 25-30% compared to our data and model predictions.

Effect of Bottom Clearance

The effects of bottom clearance on power consumption is shown in Figure 5 for a purely shear-thinning fluid (2.5% aqueous xanthan solution, $n = 0.18$) and for an elastic Boger fluid (0.08% polyacrylamide in corn syrup, $n = 0.94$) agitated by impeller $HR3$ for three different values of the bottom clearance, 10 mm, 50 mm and 85 mm respectively.

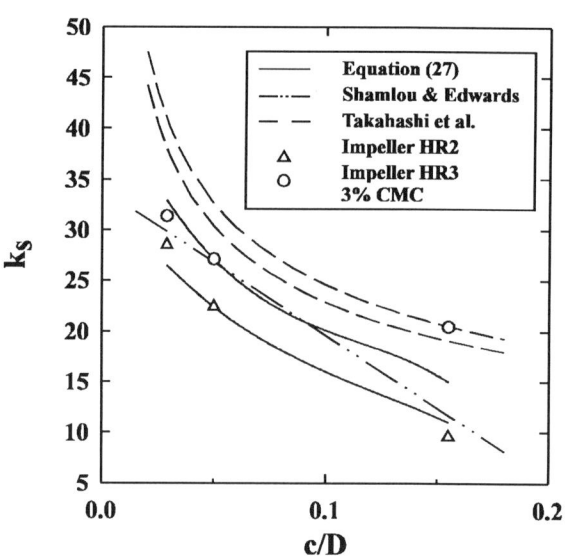

Figure 5   Effect of bottom clearance on power consumption for a shear-thinning fluid (2.5% xanthan in H_2O/glycerol and for an elastic Boger fluid (0.08% PAA in corn syrup); $c/D = 0.050$.

Included are the predictions of equation (20). Both experiments and model predictions suggest virtually no influence of the bottom clearance on power consumption (hence on k_s), at least within the range of conditions investigated in this work. This suggests little or negligible viscous dissipation in fluid between the base of the impeller and the bottom of the vessel. This observation is consistent with our previous findings for Newtonian fluids [1]. Based on the aforementioned experimental evidence, it is perhaps justified to neglect the second term on the right hand sides of equations (20) and (27). The predictions of equation (20) for the inelastic shear-thinning xanthan solution are in excellent agreement for the whole range of rotational speed. In the case of the elastic fluid (0.08% PAA in corn syrup), the predictions are satisfactory only in the lower range of rotational speed. At higher rotational speed (high Reynolds number), the power consumption is considerably increased by the fluid elasticity, as reported by Carreau et al. [1].

CONCLUSION

In this study, the effect of side wall and bottom clearances on the power consumption and the effective deformation rates produced by helical ribbon impellers in the mixing of high viscosity Newtonian and non-Newtonian systems has been investigated theoretically and experimentally. Based on the simple Couette flow representation, a theoretical framework has been developed which delineates the roles of the aforementioned two geometric parameters. Significant influence of the side wall clearance is predicted theoretically which is in line with the experimental observations. Both the effective shear rate and power consumption increase with the decreasing side wall clearance. The effect of the bottom clearance, on the other hand, is found to be negligible both theoretically and experimentally and irrespective of the non-Newtonian features of the liquid. More work is, however, required to generalize these observations over wide ranges of physical and kinematic conditions, and to other types of close-clearance impellers.

ACKNOWLEDGEMENTS

The financial supports from FCAR program of the Province of Quebec and from NSERC (for R.P. Chhabra) are gratefully acknowledged.

NOTATION

c	side wall clearance between the impeller blade to the inner vessel wall, m
d	impeller width, m
d_e	equivalent diameter of impeller, m
D	vessel diameter, m
h_c	bottom clearance, m
H	height of liquid in the vessel, m
k_s	Metzner-Otto coefficient, defined by eq. (21)
K_p	proportionality constant of the power number in laminar regime, defined by eq. (24)
m	power law parameter, $Pa.s^n$
n	power law index, -
N	impeller rotational speed, s^{-1}
Np	power number, $P/\rho N^3 d^5$
P	power, W
r	radial coordinate, m
R	vessel radius, m
Re	Reynolds number, $d^2 N \rho/\eta$
Re_g	generalized Reynolds number, $d^2 N \rho/\eta_e$
v_r	radial velocity, $m \cdot s^{-1}$
v_θ	angular velocity, $m \cdot s^{-1}$

Greek Letters

$\dot{\gamma}$	shear rate, s^{-1}
$\dot{\gamma}_{eff}$	effective deformation rate, s^{-1}
Γ	torque exerted on the vessel wall, $N \cdot m$
η	non-Newtonian viscosity, $Pa \cdot s$
μ	Newtonian viscosity, $Pa \cdot s$
ρ	liquid's density, kg/m^3
$\sigma_{\theta z}$	θz-component of stress tensor, Pa
$\sigma_{r\theta}$	$r\theta$-component of stress tensor, Pa
Ω	angular velocity of cylinder or impeller, s^{-1}

LITERATURE CITED

1. Carreau, P.J., R.P. Chhabra and J. Cheng, *AIChE J.* **39,** 1421 (1993).

2. Cheng, J. and P.J. Carreau, *Chem. Eng. Sci.*, **49**, 1965 (1994a).

3. Harnby, N., Edwards, M.F. and A.W. Nienow, *Mixing in the Process Industries*, 2nd ed., Butterworth, London (1992).

4. Metzner, A.B. and R.E. Otto, *AIChE J.* **3**. 3 (1957).

5. Tatterson, G.B., *Fluid Mixing and Gas Dispersion in Agitated Tank*, McGraw Hill, New York (1991).

6. Ulbrecht, J.J. and P.J. Carreau, *Mixing of Liquids by Non-Newtonian Liquid* in *Mixing of Liquids by Mechanical Agitation*, J.J. Ulbrecht and G.K. Patterson (Eds.), Gordon and Breach Publishers, New York (1985).

7. Kappel M., *Int. Chem. Eng.*, **4**, 571 (1979).

8. Nagata, S., *Mixing: Principles and Applications*, Kodansha, Wiley, Tokyo (1975).

9. Oldshue J.Y., *Fluid Mixing Technology*, McGraw Hill, New York (1983).

10. Hall, K.R. and J.C Godfrey, *Trans. Instn. Chem. Eng.*, **48**, T201 (1970).

11. Takahashi, K., T. Yokota and H. Konno, *J. Chem. Eng. Japan*, **17**, 657 (1984).

12. Ayazi Shamlou, P. and M.F. Edwards, *Chem. Eng. Sci.*, **40**, 1773 (1985).

13. Bourne, J.R., and H. Butler, *Chem. Eng. J.* **47**, 263 (1969).

14. Bird, R.B., R.C. Armstrong and O. Hassager, *Dynamics of Polymeric Liquids, Vol.1, Fluid Mechanics*, John Wiley & Sons, New York (1987).

15. Cheng, J. and P.J. Carreau, *Can. J. Chem. Eng.*, **72**, 418 (1994b).

16. Chavan V.V., and J.J. Ulbrecht, *Ind. Eng. Chem. Proc. Des. Develop.* **12**, 472 (1973).

17. Yap, C.Y., W.I. Patterson and P.J. Carreau, *AIChE J.*, **25**, 516 (1979).

Study of Micromixing in a Liquid-Solid Suspension in a Stirred Reactor

P. Guichardon, L. Falk, M.C. Fournier and J. Villermaux
Laboratoire des Sciences du Génie Chimique CNRS, Ecole Nationale Supérieure des Industries Chimiques INPL, 1 Rue Grandville, BP 451 - 54001 NANCY Cedex, France

1. Introduction

The motivation of the present study is precipitation. It is well known that micromixing effects may play a significant role in precipitation. The first stage of this process is generally the contacting of two liquid reactants A and B either in a single jet or in a double jet operation. A and B react quasi instantaneously and yield the insoluble product P, first in a supersaturated form which further nucleates, grows and finally gives rise to solid particles. Micromixing at the feed point of reactants has two adverse effects. On one hand it favors contact between A and B and increases supersaturation. On the other hand, micromixing smoothes out local concentration gradients and causes supersaturation to decrease. The rate of nucleation is a very stiff function of supersaturation with some kind of threshold effect whereas the rate of growth of particles increases more smoothly with supersaturation.

Therefore, micromixing may be expected to exert a complex influence on particle size distribution and on crystal nature and morphology.

As local micromixing efficiency depends on the rate of power dissipation per unit mass ε, in the vicinity of the injection point, it is essential to know the distribution pattern of this quantity in the precipitation vessel in order to select the best position of the feed pipes and to control the particle size distribution.

Data on ε distribution, mostly expressed as $\phi = \varepsilon / \bar{\varepsilon}$ where $\bar{\varepsilon}$ is the average power dissipation per unit mass, can be found in the literature for clear liquids [1]. Data on liquid-solid suspensions are more scarce, especially as far as small particles — a few tens of microns — are concerned [2]. Local measurements of micromixing efficiency by means of a chemical test reaction yield useful information about ε, as most characteristic micromixing times in the viscous subrange are proportional to $(\nu/\varepsilon)^{1/2}$. It is therefore desirable to extend such measurements, which are common in clear liquids, to suspensions of solid particles encountered in precipitation processes. A few determinations of this kind, mainly in the case of fiber dispersions, are reported in the recent literature [3] [4].

Having developed our new iodide-iodate reaction, we performed a few experiments in dilute suspensions of glass beads whose preliminary results are reported in reference [5]. Actually these results were somewhat surprising as they seemed to indicate that micromixing efficiency was enhanced by the presence of suspended solids without any increase of the average power dissipation. This puzzling result might be explained by a redistribution of ε under the influence of solid particles but it was nevertheless deserving of further confirmation. This is the aim of the present paper where we first recall the principle of the experimental procedure and then report new results which show that the presence of solid particles has only a weak effect, if any, on micromixing efficiency. The previous results were probably biased by a loss of iodine, whose mass balance has to be carefully controlled in order to obtain reliable results.

2. Experimental procedure

Chemical Test-System

The chemical system used in this study is the iodide-iodate reaction coupled with a neutralization reaction (1)

$$H_2BO_3^- + H^+ \rightleftarrows H_3BO_3 \quad (1)$$

$$6H^+ + 5I^- + IO_3^- \rightleftarrows 3I_2 + 3H_2O \quad (2)$$

$$I_2 + I^- \rightleftarrows I_3^- \quad (3)$$

The role of boric acid is to keep the pH above 9 where reaction (2) does not occur to a significant extent. In the presence of an excess of acid, iodine is released and immediately complexed by iodide to yield triiodide which is monitored by spectrophotometry at 353 nm.

Under typical conditions the I^-/IO_3^- mixture is stoichiometric so that one can introduce "potential iodine" as a pseudo-compound whose initial concentration is

$$C'_{co} = \frac{3}{5}\left(C'_{I^-}\right)_o = \frac{1}{5}\left(C'_{IO_3^-}\right)_o$$

The reacting system can then be written in the simpler form of two parallel competing reactions :

$$\begin{cases} A + B \rightarrow R & (4) \\ C + 2B \rightarrow S & (5) \end{cases}$$

where A and B and S respectively represent $H_2BO_3^-$, H^+ and $I_2 + I_3^-$

The concentrations before mixing used for the experiments are :

$C'_{H_2BO_3^-} = 0.089$ M $C'_{I^-} = 0.01166$ M

$C'_{IO_3^-} = 0.0023$ M $C'_{H^+} = 4$ N

The kinetics of reaction (2) — the classical Dushman reaction — were carefully reinvestigated under operating conditions. The rate of reaction was finally found to be

$$r = k\, C_{H^+}^2\, C_{I^-}^2 \cdot C_{IO_3^-} \quad (6)$$

where k is a function of ionic strength I

T = 293 K

I<0.16 M $\log_{10} k = 9.281 - 3.664\sqrt{I}$

0.16<I<2M $\log_{10} k = 8.383 - 1.5112\sqrt{I} + 0.23689\, I$

with concentrations in mole/liter.

Reactant B is feed into a mixture of A and C in excess. Under perfect mixing conditions, no S should appear. The amount of S — here iodine — produced by the reaction is thus a measure of segregation, which results in a local excess of B in the mixing "plume". After completion of the reaction, a sample of the solution is taken out, filtered in order to remove the solid particles and analyzed in the spectrophotometer. The segregation index is calculated from the amount of I_3^- formed as

$$X_s = \frac{n_{I_2} + n_{I_3^-}}{(n_{H^+})_o} \left\{ 2 + \frac{(n_{H_2BO_3^-})_o}{3(n_{IO_3^-})_o} \right\} \quad (7)$$

where n stands for mole number and subscript o for initial conditions.

The results are expressed in terms of the micromixedness ratio $\alpha = \frac{1-X_s}{X_s}$ which may be interpreted as the ratio of the perfectly micromixed volume fraction over the segregated volume fraction in the mixing plume. Therefore, the greater α, the better the micromixing efficiency.

Reactor

The reactor is a 20 liters tank in standard configuration equipped with 4 baffles and a 6 bladed Rushton turbine or a TTP Mixel propeller.

Figure 1 shows the main characteristics of the tank and the 3 different locations of feed points :

<1> above the stirrer
<2> in the plane of the stirrer
<4> in the discharge flow of the turbine

The tank is closed with a lid in order to avoid entrainment of air bubbles into the liquid, especially at high rotation speed.

The diameter of the feed pipe is 4 mm. The shaft of the impeller is equipped with a torque meter to measure the power consumption.

Solid particles

Glass beads of different sizes were used.

Small size : d_p = 27 µm
Medium size : d_p = 201 µm
Large size : d_p = 1250 µm

These are mass-average values. Figure 2 shows the particle size distribution of the small and medium size samples. The density of glass particles is ρ_s = 2500 kg.m^{-3}. Before use, the particles were washed in sulfuric acid medium in order to clean-up the surface and to prevent possible adsorption. The beads were then rinsed and dried.

Determination of the just-suspended speed N_{js}

N_{js} was estimated from the standard Zwietering's correlation [6], including for large particles suspended with the TTP Mixel impeller although no data were available for this specific system. Actually, we checked visually the complete suspension of glass beads in this case. The calcultated N_{js} values are given below :

Particles	Impeller	N_{js} (s^{-1}) for several solid mass fraction			
		1 %	2 %	4 %	6 %
d_p = 27 µm	Rushton	5.7	6.2	6.8	7.2
d_p = 201 µm	Rushton	8.5	9.2	10.1	10.7
d_p = 1250 µm	TTP Mixel	12.2	13.4	14.6	15.5

According to literature data [7], the suspension can be considered as quasi homogeneous within the whole reactor for the smaller particles but probably not for medium and large size particles.

Mean power dissipation

Several experiments were carried out in order to determine the mean power dissipation in the tank as a function of stirring speed and solid mass concentration, mainly with small particles, $d_p = 27$ µm. Figure 3 clearly shows that there is no influence of solids on mean power dissipation even up to a relatively high mass concentration of 16 %. This result is in agreement with data obtained by other authors [4] [8].

Nevertheless according to Geisler [2], the distribution of local power dissipation in stirred tank can be slightly modified as a function of mass concentration and particle size in liquid-solid suspension. Changes are more pronounced near the impeller whereas they are almost negligible in the rest of the tank. For our micromixing study, several injection points were thus selected at different positions, close to the impeller and in the mean flow, in order to investigate the effect of solid particles on local power dissipation.

Precautions for using the test reaction in the presence of solid particles

The determination of micromixing efficiency relies on measuring the amount of iodine released by the reaction. The better the micromixing, the smaller this amount. Consequently, all phenomena causing a loss of iodine before analysis may lead to the wrong conclusion of enhanced micromixing. Two sources of iodine loss were identified : adsorption onto solid particles causing a depletion of iodine in the liquid phase and evaporation during filtration of the suspension. Adsorption may be suspected to occur especially onto smaller particles which exhibit a large surface area despite the low affinity of glass for inorganic compounds. After carefully checking the iodine mass balance, it was deduced that losses by evaporation during filtration were probably responsible for the apparent increase of micromixing efficiency reported in our preliminary experiments [5]. In the case of the larger particles used in this study, filtration was not required and solids could be separated from the suspension by simple settling. There is neither adsorption nor evaporation in this case and the results can be directly exploited. The data reported below with smaller particles were as far as possible corrected for these spurious effects.

3. Results

A small amount of acid (1 ml) was injected into the tank slowly enough (200 s) to avoid macromixing phenomena. It was checked that the critical feed time in the presence of small particles was not very different from that in clear liquid.

Figure 4 shows a plot of the micromixedness ratio as a function of mass concentration of small particles ($d_p = 27$ µm) for several locations of injection points. Due to the scatter of experimental data, no systematic effect of glass beads on micromixing efficiency can be detected. This is true whatever the location of the injection point. This last result tends to show that the local power dissipation in stirred tank is not significantly changed by the presence of solid to affect micromixing phenomena.

Similar results are presented in figure 5a, for medium size particles ($d_p = 201$ µm) : the presence of solids up to 5 % in mass has no significant influence on micromixing and no difference can be detected between injection points <1> and <2>. Figure 5b shows results obtained with large particles ($d_p = 1250$ µm) and TTP Mixel propeller : here again micromixing is not significantly modified by solid particles.

4. Discussion

Substantial turbulence modification in particulate flows has been reported in several experimental studies. Information exists in the literature on the reduction or increase of the turbulence intensity caused by the presence of a dispersed phase in a fluid. According to Gore et al. [9] who compiled most of the available data, small particles reduce the turbulence intensity of the flow, while larger particles increase it.

Previous studies [5] have shown that in the range of reactants concentrations used here, the micromixedness ratio varies with power dissipation according to :

$$\alpha \approx \varepsilon^{0.14} \qquad (8)$$

Due to the low value of this exponent, small changes in ε caused by the presence of solid particles in dilute suspensions have a negligible effect on α.

This explains the results obtained in this study.

Conclusion

After correction for spurious effects, the iodide-iodate test reaction was found useful for studying micromixing efficiency in dilute suspension of glass beads in stirred tanks.

In the range 20 µm < d_p < 1300 µm, 0 % < C_s (by weight) < 5 %, glass beads in a liquid-solid suspension have no influence on average stirring power dissipation and a negligible effect on micromixing intensity. The interesting consequence of this observation is that data for clear liquids may be used for the design of precipitation reactors, at least for dilute suspensions.

Future work should be devoted to the study of more concentrated suspensions and take into account both the local distribution of power dissipation and of solid concentration.

Notation

C	concentration, mol.l^{-1}
C_s	solid mass fraction, %
d_p	diameter of glass particles
I	ionic strength, mol.l^{-1}
k	kinetic constant, M^{-4}.s^{-1}
N	stirring speed, s^{-1}
N_{js}	just suspended speed, s^{-1}
n	mole number, mol
P_s	power dissipation in liquid solid suspension, W
r	reaction rate, M.s^{-1}
T	temperature, K
X_s	segregation index
α	micromixedness ratio
ϕ	$\varepsilon/\overline{\varepsilon}$ factor
$\underline{\varepsilon}$	local rate of energy dissipation, W.kg^{-1}
ε	average rate of energy dissipation, W.kg^{-1}
ν	kinetic viscosity, m^2.s^{-1}
ρ_s	density of glass particles, kg.m^{-3}

References

[1] Laufhütte H.D., Mersmann A., "Local energy dissipation in agitated turbulent fluids and its significance for the design of stirring equipment", Chem. Eng. Technol., 10, p. 56-63, 1987

[2] Geisler R.K., "Fluiddynamik und Leistungseintrag in turbulent gerührten Suspensionen". PhD Thesis, Technical University Munich, 1991

[3] Bennington C.P.J., Bourne J.R., "Effect of suspended fibres on macromixing and micromixing in a stirred tank reactor", Chem. Eng. Comm., vol. 92, p. 183-197, 1990

[4] Bennington C.P.J., Thangavel V.K., "The use of a mixing sensitive chemical reaction for the study of pulp fibre suspension mixing", Can. J. Chem. Eng., vol. 71, p. 667, October 1993

[5] Villermaux J., Falk L., Fournier M.C., "Potential use of a new parallel reaction system to characterize micromixing in stirred reactors", AIChE Symposium series, n° 299, vol. 80, p. 50-54, 1994

[6] Zwietering Th.N., Chem. Eng. Sci., $\underline{8}$, p. 244-253, 1958

[7] Baldi G., Conti R., Gianetto A., AIChE J., vol. 27, n° 6, p. 1017, November 1981

[8] Drewer G.R., Ahmed N., Jameson G.J., "Suspension of high concentration solids in mechanically stirred vessels", 8th European Conf. on Mixing, Symposium series n° 136, p. 41, 21-23 Sept. 94

[9] Gore R.A., Growe C.T., "Effect of particle size on modulating turbulent intensity", Int. J. Multiphase Flow, $\underline{15}$, p. 279-285, 1989

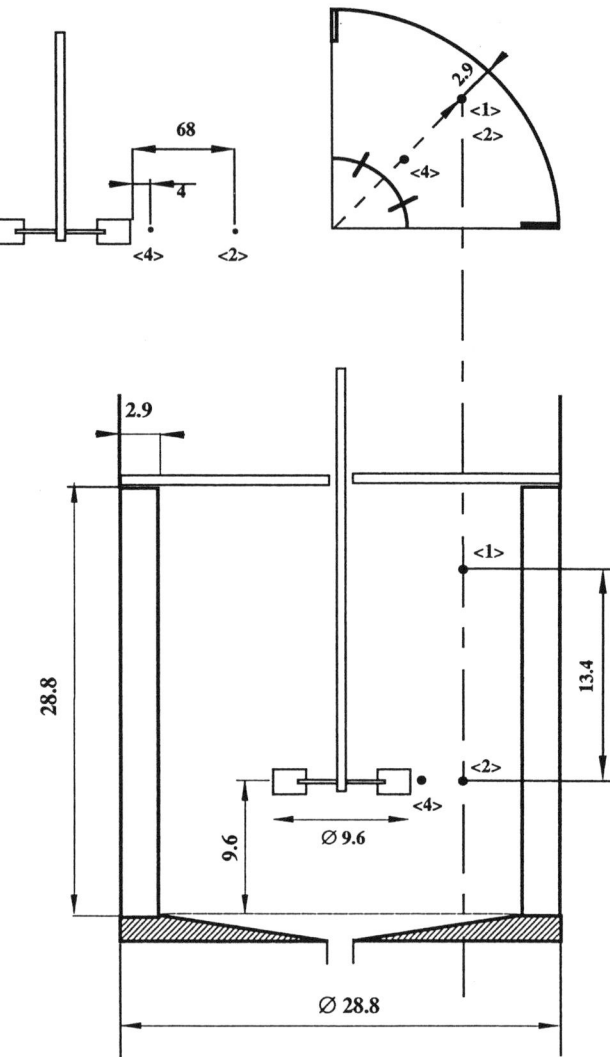

Figure 1 : Sketch of the stirred tank reactor and position of injection <1>, <2>, <4> (dimensions in mm)

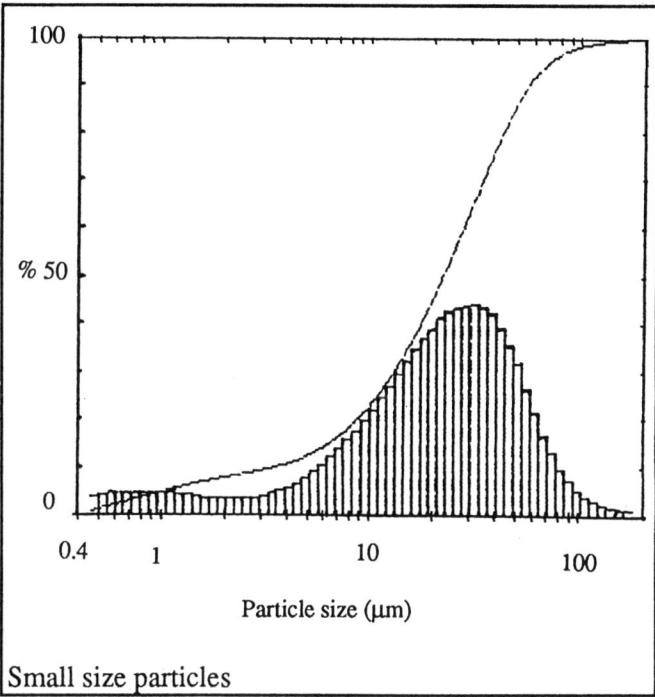

Small size particles

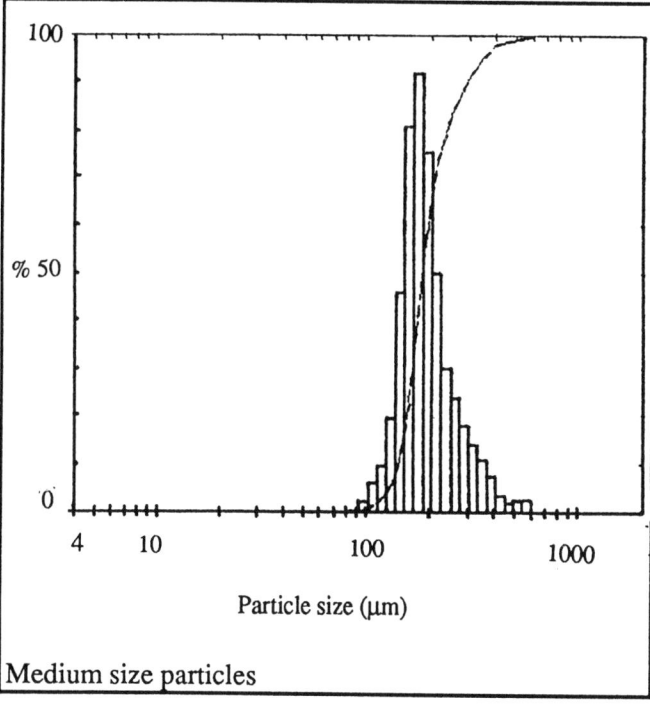

Medium size particles

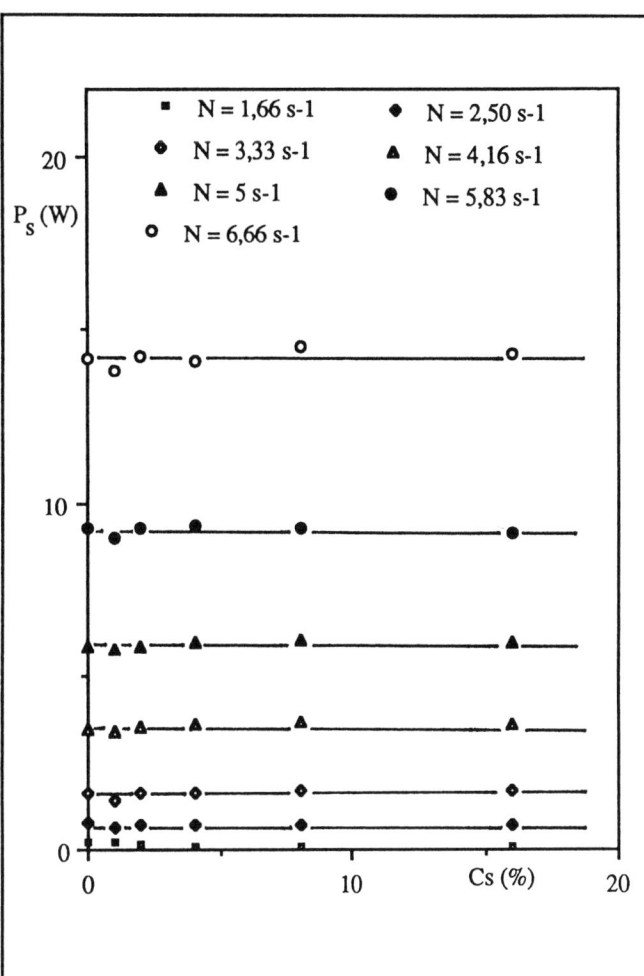

Figure 3: Power dissipation in liquid-solid suspensions for the small size particles

Figure 2: Particle size distributions of the small and medium sizes particles

130   Industrial Mixing Fundamentals with Applications

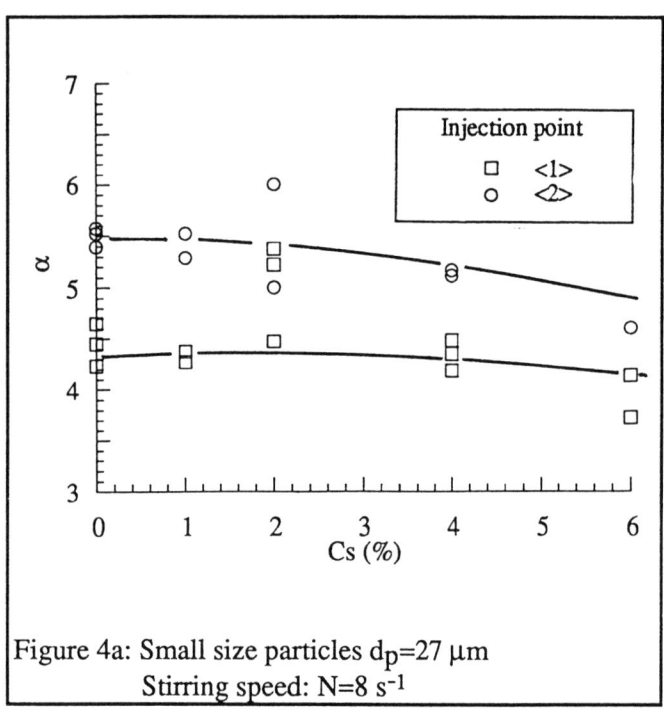

Figure 4a: Small size particles $d_p=27$ μm
Stirring speed: $N=8$ s^{-1}

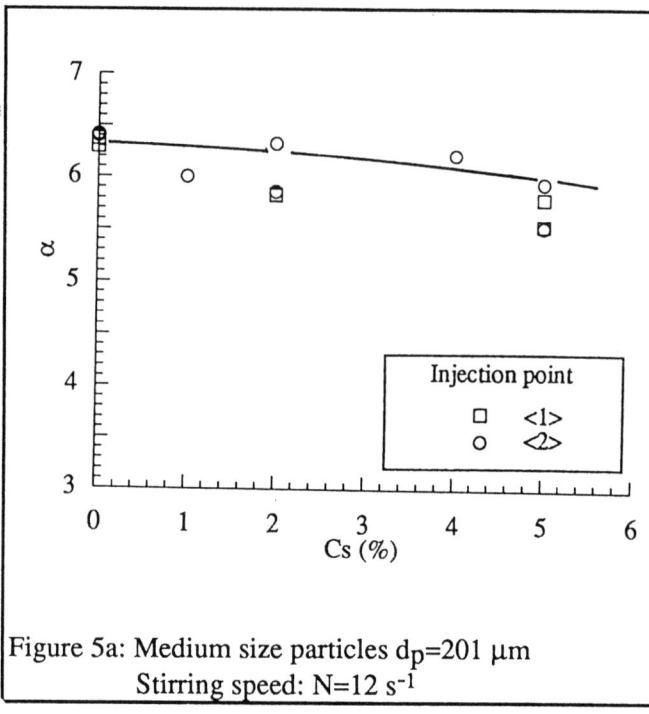

Figure 5a: Medium size particles $d_p=201$ μm
Stirring speed: $N=12$ s^{-1}

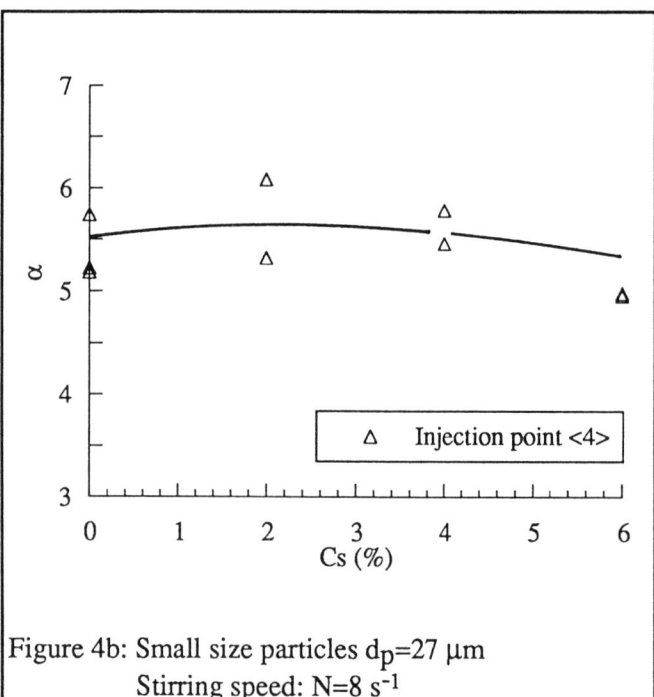

Figure 4b: Small size particles $d_p=27$ μm
Stirring speed: $N=8$ s^{-1}

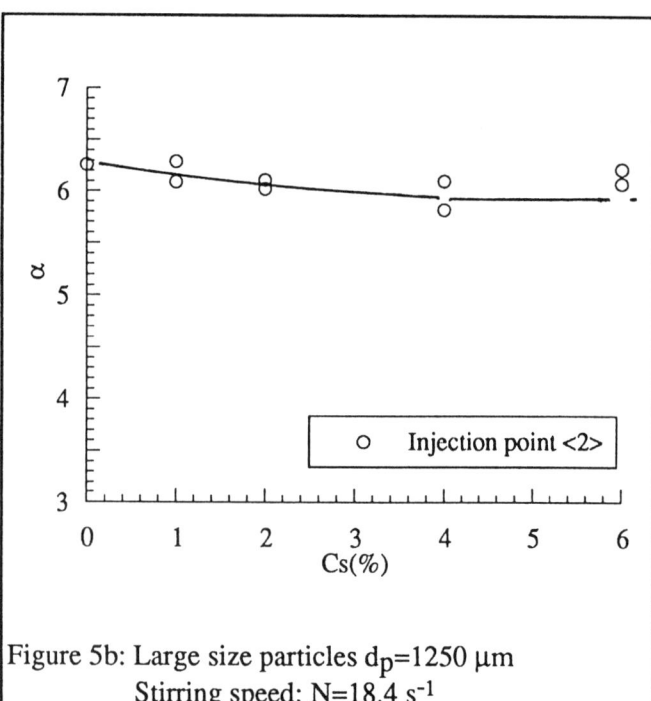

Figure 5b: Large size particles $d_p=1250$ μm
Stirring speed: $N=18.4$ s^{-1}

Figure 4: Experimental evolution of micromixedness ratio α against solid mass fraction for the small size particles

Figure 5: Experimental evolution of micromixedness ratio α against solid mass fraction for the medium and large sizes particles

An Experimental Investigation of Solids Suspension at High Solids Loadings in Mechanically Agitated Vessels

Naimul H. Chowdhury, W. Roy Penney
University of Arkansas, Fayetteville, AK 72701

Kevin J. Myers
University of Dayton, Dayton, OH 45469

Julian B. Fasano
Chemineer, Inc., 5870 Poe Avenue, Dayton, OH 45401

An experimental investigation has been conducted to determine the effect of vessel size, D/T and solid properties on solids suspension at high solids loadings for 4BP, 6BD & HE-3 impellers. Five solids were suspended in water in 19, 66, 167 & 600 l vessels. At high solids loadings the just suspended speed increases rapidly with solids loading. The best correlation of the data for various solids in several vessels was obtained by plotting a relative just-suspended speed (ratioed to the speed at zero solids loading) vs settled volume fraction solids. At high solids loadings larger and more angular solids require a higher N_{jsr} than do smaller more rounded solids. The increase in N_{js} with increasing solids loading is much less pronounced for high D/T (up to 0.64) than for low D/T (down to 0.25). Hysterisis has been quantified for one solid and suspension collapse has been observed at high solids loadings.

INTRODUCTION

High solids loadings are a fact of life in many liquid-solid chemical processes, as they are used to maximize a plants output per volume of processing equipment. With respect to agitators and other pieces of processing equipment, there are economic limits to solids loadings, and it is necessary to understand the behavior of high solids loadings in order to define this economic limit. In the vicinity of these "economic" loadings, agitator size increases dramatically with an increase in solids loading. Very often a change of only several percent solids can double the size of the agitator required.

The present paper is an interim report of a comprehensive investigation of solids suspension in agitated vessels at high solids loadings. The findings thus far indicate that, for all impeller styles, complex interactions exist between solids characteristics, solids loading, D/T and vessel size. The present paper discusses some of those interactions:

(1) the effect of solids loading on just-suspended and maintain-suspend speed in various size vessels for a given solid,

(2) the speed differential due to the hysterisis effect of achieving-suspension speed vs maintaining-suspension speed.

(3) the effect of solid properties on N_{js} for four different solids at various solids loadings, vessel sizes and D/T,

(4) the effect of D/T on N_{js} for two solids at X_w = 0.05 and 0.40.

LITERATURE REVIEW

For a general review of the solids suspension literature the reader is referred to the work of Nienow [1]. The focus of the present paper is high solids loadings, up to the point where 90 % of the liquid plus solid (i.e., batch) volume is filled with settled solids. Drewer et. al [2] and Buurman [3] have recently studied high solids concentrations. Buurman [3] obtained experimental data with 45 degree pitched-blade impellers (4BP) with $D/T = 0.4$ using several solids suspended in water, one solid in octane and one in butanol in a 10 l vessel and 3 mm silica spheres in water in 10, 80 & 4500 l vessels. The data were plotted in the form of $[(\rho/\Delta\rho)^3(N_{js}D/g)^3D/d_p]^{1/6} = K_{js}$ vs volume fraction solids (X_v). K_{js}, which is proportional to N_{js} for a given solid and vessel, increased gradually with increasing X_v until an X_v was reached where K_{js} started rising rapidly with increasing X_v. For the larger solids (2 to 3 mm) in 10 & 80 l vessels the rapid rise started at $X_v = 0.25$ to 0.3 but for the smaller solids in the small vessels and for the 3 mm solids in the 4500 l vessel the rapid rise of K_{js} did not start until $X_v > 0.4$ to 0.5. At high X_v the rise in N_{js} was very steep, e.g., for the 3 mm silica spheres in the 4500 l vessel, N_{js} increased fourfold as X_v increased from 0.4 to 0.5.

Drewer et al [2] used a 6.3 l vessel and tested six-blade disk impellers (6BD), 4BP and Marine Propellers (MP) all with $D/T = 1/3$. Four sizes of glass beads and one resin were suspended in water. X_v was varied up to 0.49. The suspension data were plotted as impeller power (P) vs X_v. N_p varied with X_v. For the MP, N_p initially increased from 0.82 at $X_v = 0$ to about 1.0 at $X_v = 0.25$ and then remained about constant. For the 4BP, N_p was about constant at 1.35. For the 6BD, N_p decreased from 5.2 to about 4 at $X_v = 0.5$ for the larger solids but decreased to about 2.5 for the smallest solids. For all solids and all impellers, a rapid rise in P started occurring about $X_v = 0.4$. Baffle elevation off the vessel bottom was investigated. The MP gives best performance with full length baffles; however, for the 6BD, an increase of off-bottom baffle clearance dramatically decreased P at high X_v - a baffle elevation of 0.4T gave a 4-fold reduction in P at $X_v = 0.3$. Drewer et al [2] also observed hysterisis for $X_v > 0.25$, i.e., "the power required to initiate motion, i.e. to "un-sand" the impeller was often greater than the power required for off-bottom suspension" and suspension collapse, i.e., "a point is reached at which a small decrease in power resulted in complete collapse of the suspension".

EXPERIMENTAL APPARATUS

A schematic of the experimental apparatus is presented as Figure 1. Four geometrically similar, flat-bottom Plexiglas vessels with diameters of 28.9, 43.8, 59.9 and 91.4 cm (volumes = 19, 66, 167 and 600 l) were used. Two photoflood lights along with a viewing mirror placed underneath the vessel allowed visual views of the vessel side and bottom. A protector plate was placed on top of the baffles when operating at the highest impeller speeds to prevent excessive surface aeration.

Three different impellers (a six-bladed disk turbine [6BD], a four-bladed, 45 degree pitched-blade turbine [4BP] and a Chemineer HE-3) were tested. The HE-3 impellers were the standard Chemineer geometry; the 4BP impellers had $W/D = 0.2$ and the 6BD impellers had $W/D = 0.2$ and $L/D = 0.25$. Impeller diameters were varied, for any particular vessel, to investigate the effect of D/T on the just-suspended condition.

EXPERIMENTAL SOLIDS

Five different solids were suspended in water. The characteristics of the solids are given in Table 1. Three plastic pellets, a sand and an alumina were tested. The size varied from 600 to 3600 microns, the specific gravities varied from 1.18 to 3.5 and the settling velocities varied from 7.7 to 21 cm/sec.

EXPERIMENTAL PROCEDURE

The just-suspended condition was determined as follows:

(1) When the solids moved to the bottom of the vessel and then were immediately swept off the vessel bottom, the just-suspended condition occurred at the impeller speed where solids rested on the vessel bottom for a maximum of one to two seconds.

(2) At higher solids loadings solids would sometimes move jerkily along the vessel bottom at intervals of one to two seconds, but the solids were never completely suspended off the vessel bottom. At this condition the just-suspended speed was the speed which gave off-bottom suspension at maximum intervals of five to ten seconds.

(3) Under high solids loadings a hysterisis effect existed, sometimes in conjunction with suspension collapse, as noted by Drewer et al [2]. Under hysterisis conditions a higher speed was required to unlock and

Table 1. Properties of the Experimental Solids

Solid	Shape	Major Dia. cm	Minor Dia. cm	Length cm	S.G.	U_t in water cm/sec
LP	Oval cyl.	0.360	0.224	0.287	1.35	11
SP	Round cyl.	0.201		0.305	2.0	21
S	Grains	0.06			2.54	10
AL	Crushed	0.06			3.50	12
LA	Pellets	0.32			1.18	7.7

Nomenclature: LP = Large polyethylene terephthalate
SP = Small polyethylene terephthalate
S = Sand
AL = Alumina
LA = Large acrylic beads, rectangular & oval cylinders

initially suspend (i.e., achieve-suspension) the static bed of solids than was needed to maintain (maintain-suspension) the just-suspended condition; the just-suspended speed was taken as the achieve-suspension condition. Under hysterisis conditions sometimes a small decrease in speed resulted in a complete collapse of the suspension. Invariably the collapse speed was lower than the achieve-suspension speed.

CORRELATING PARAMETERS TO INCLUDE EFFECT OF SOLIDS LOADING

Previous investigators have used weight fraction solids (X_w) and volume fraction solids (X_v) as parameters to correlate the effect of solids loading. Another parameter - the fractional settled height of solids ($F_s = H_{ss}/Z$) - is used as a correlating parameter here. The rational for using F_s rather than either X_w or X_v is that F_s is expected to provide an upper bound of N_{js} of infinity at $F_s = 1$; i.e., the solids are impossible to suspend when the vessel is completely filled with settled solids. Neither X_w or X_v give a common upper bound for varying solid and liquid densities. For comparison purposes, relationships are needed among X_s, X_v, F_s and ε (ε is the volume fraction of liquid in the settled solids bed, i.e., the void fraction in the bed):

$$X_v = 1/[1 + (\rho_s/\rho)(1/X_w - 1)]$$

$$F_s = X_v/(1-\varepsilon)$$

$$F_s = [1/(1-\varepsilon)]/[1 + (\rho_s/\rho)(1/X_w - 1)]$$

EFFECT OF SOLIDS LOADING ON N_{JS} FOR A PARTICULAR SOLID WITH AN HE-3 IMPELLER - INCLUDING HYSTERISIS (Figure 2)

A series of experiments were conducted with the Acrylic pellets aimed at determining the relationship between N_{js} and solids loading for three vessel sizes (V = 19 l, 167 l & 600 l) for a particular solid. For these experiments only one HE-3 impeller (D/T = 0.34) was tested in each vessel. Figure 2 presents all the data for all three vessels. The data are plotted as relative just-suspend speed (N_{jsr}), which is a ratio of N_{js} at a given solids loading to N_{js} at zero solids loading. Any solids loading can be chosen as the basis for setting the reference N_{js} but selecting the reference N_{js} at $X_w = 0$ is attractive because it is not yet known which parameters, or combination of parameters, relating to solids loading (e.g., X_w, X_v or F_s etc), will eventually best correlate the effect of solids loading. Choosing a reference N_{js} at zero solids loading has the advantage of fixing the reference speed provided any correlating parameter chosen is zero at zero solids loading.

The open symbols plot the achieve-suspension data and the solid symbols plot the maintain-suspension data. For small vessels and/or low F_s, the achieve-suspension speed and maintain-suspension speed are the same. When the achieve-suspension speed and the maintain-suspension speed were the same, to avoid a "busy", cluttered plot, the achieve-suspension symbols are used for plotting. Pertinent observations from the experimental investigations and the data presented in Figure 2 are:

(1) There is a definite effect of scale on the hysterisis effect. The magnitude of the hysterisis effect is indicated by the difference between (a) the achieve-suspension speed and (b) the maintain-suspension speed. The Acrylic pellets exhibited negligible hysteresis at solids loading below about $F_s = 0.29$ ($X_w = 0.2$) for all the vessels. But there is significant hysteresis for the 167 l and 600 l vessels for $F_s > 0.36$ ($X_w > 0.25$). In fact at $F_s = 0.52$ ($X_w = 0.35$) the hysterisis effect for the 600 l vessel gives a 20% higher speed needed to achieve suspension than the speed needed to maintain suspension. At higher F_s the difference is expected to become even greater.

(2) The data point at $F_s = 0.67$ ($X_w = 0.45$) for the 19 l vessel was obtained with some difficulty. At this solids loading in the 19 l vessel, N_{js} starts to increase rapidly with an increase in solids loading. At these conditions the just-suspended speed is difficult to determine visually. For the 167 l vessel at $F_s = 0.64$ ($X_w = 0.4$) and for the 600 l vessel at $F_s = 0.76$ ($X_w = 0.5$) the drive units did not have adequate torque capability to achieve suspension. However, by gradually lowering the impeller into the static bed of solids the maintain-suspension N_{js} could be determined at $F_s = 0.72$ ($X_w = 0.475$) for the 600 l vessel.

(3) The shape of the curve of N_{jsr} vs F_s is the same general shape as for all the data of this investigation: N_{js} rises rapidly with F_s until F_s is about 0.05; above $F_s = 0.05$ the rate of increase becomes more gradual and then as $F_s > 0.6$, N_{jsr} starts increasing rapidly with increased solids loading.

(4) For these acrylic beads, the correlation of N_{jsr} vs F_s gives a reasonable correlation of the data for the three vessel sizes. Correlations vs either X_w or X_v would correlate all the data equally well.

EFFECT OF SOLIDS TYPE AND SOLIDS LOADING ON JUST-SUSPENDED SPEED
(Figures 3-8)

The following range of experimental conditions was investigated to determine the effect of solids loading on the just-suspended speed for three impellers and a variety of solids: Impellers: 6BD, 4BP & HE-3; Solids: LP, SP, S, AL; D/T: 0.36 & 0.52; Vessel Volumes: V = 66 & V = 167 (for LP only).

Figures 3 thru 8 present the data plotted as N_{jsr} vs F_s. The values of the just-suspended speed (N_{js}) at $F_s = 0$, i.e. the reference speed, are given in RPM below the legend identifying the individual solids in the caption immediately above each of Figures 3-8. For all the data presented in Figures 3-8, F_s was determined experimentally at each data point.

Figures 3 presents data for an HE-3 impeller for V = 66 & 167, respectively, for D/T = 0.36. Figure 4 presents analogous data for D/T = 0.52. Figure 5 & 6 and 7 & 8 present analogous data for 4BP and 6BD, respectively. Pertinent observations from the experimental investigations and the data presented in Figures 3-8 are:

(1) The scatter in the data at high F_s is a result of practical difficulties of obtaining accurate and precise data at high solids loadings. The speeds are so high at the highest solids loadings that degradation of solids and impellers occur; thus, one needs to obtain visual determination of N_{js} quickly. Our ongoing studies are aimed at the acquisition of more consistent data. Videotaping with replay in slow motion is being used as a tool to improve both accuracy and precision of the just-suspended speed.

(2) As for the acrylic pellets, Figures 3-8 typically show a rapid increase of N_{jsr} vs F_s at low F_s, then level off at intermediate loadings and eventually show a rapid increase of N_{jsr} with F_s at high F_s.

(3) At F_s approaches 0.9, the solids become almost impossible to suspend with the suspending speed being 7 to 8 times higher at $F_s = 0.9$ than at $F_s = 0$. All the agitator drives we have used become either torque or speed limited at the highest values of F_s tested.

(4) Although the data are somewhat scattered, N_{jsr} at higher solids loadings is higher for the plastic pellets than for the sand and alumina. This is perhaps a result of (a) the more free-flowing nature of the sand and alumina compared to the plastic pellets [the angle of repose is somewhat greater for the plastic pellets than for the sand and alumina] and (b) the larger size of the pellets than the sand and alumina. Buurman [3] found that his correlating parameter was higher for large solids than for small solids at high X_v.

(5) The effect of increased solids loading on the increase in N_{jsr} in the intermediate range of $F_s = 0.2$ to $F_s = 0.7$ is much less for a D/T of 0.52 than for a D/T of 0.36. This finding will be further discussed next when the data taken specifically to determine the effect of D/T on N_{js} are discussed.

(6) The effect of solid properties on N_{js} are reasonably well correlated by plotting N_{jsr} vs F_s; however, Figures 3-8 indicate a need to develop a more accurate

EFFECT OF D/T ON N_{JS} FOR HE-3 IMPELLER WITH SAND AND AN ION-EXCHANGE RESIN IN A 19 L VESSEL AT SOLIDS LOADINGS OF 5 & 40 w% (F_s ABOUT 0.07 & 0.60)
(Figures 9 & 10)

Experiments were conducted with sand and ion-exchange resins aimed at determining the relative effect of D/T on N_{js} as solids loading varied. Figure 9 presents sand suspension data for an HE-3 impeller in a 19 l vessel and Figure 10 presents resin suspension data for an HE-3 impeller in a 19 l vessel. Pertinent observations from the experimental investigations and the data presented in Figures 9 & 10 are:

(1) At low D/T the effect of solids loading on N_{js} is very pronounced, but as D/T increases the effect of solids loading becomes much less and, in fact, for the sand at a D/T of 0.63, N_{js} is lower for X_s = 40 w% than for 5 w%. This behavior is also very evident by comparing the data for D/T = 0.52 with the data for D/T = 0.36 by inspection of Figures 3-8. In the intermediate range of F_s from 0.2 to 0.6, the increase of N_{js} with an increase in F_s is much less pronounced for D/T = 0.52 than for 0.36 for all three impellers tested. This behavior perhaps indicates that there is a potential economic advantage for the use of higher D/T at high solids loadings. Our future work will investigate this possibility.

(2) The relative of effect of D/T on N_{js} is the same for the both solids, except that the resin data do not actually show a crossing of the curves for X_w = 0.05 and for X_w = 0.4 at the highest D/T tested of 0.64.

ACKNOWLEDGEMENTS

We gratefully acknowledge the help and cooperation of the following people: G. Richard Goodley for facilitating the donation of the polyester flake by the Du Pont Company; Thomas J. Dillon for facilitating the donation of the alumina by Alcoa; and Arthur W. Etchells of Du Pont for suggesting that the settled volume fraction of solids (F_s) might be an appropriate correlating parameter for the effect of solids loading.

REFERENCES

1. Nienow, A. W., "The Suspension of Solid Particles". Chapter 16 in Mixing in the Process Industries, 2nd ed., N. Harnby, M. F. Edwards and A.W. Nienow, (Eds), Butterworths, London (1992).

2. Drewer, G. R., Ahmed, N. and Jameson, G. J., "Suspension of High concentration solids in Mechanically Stirred Vessels", I.CHEM.E. SYMPOSIUM SERIES NO. 136, p. 41-48 (1994).

3. Buurman, C., "Stirring of Concentrated Slurries: A Semi-empirical Model for Complete Suspension at High Solids Concentrations and 5 m³ Verification Experiments", I.CHEM.E. SYMPOSIUM SERIES NO. 121, p. 343-350 (1990).

NOMENCLATURE

B	-	Baffle width, m
C	-	Impeller height (i.e., clearance) off vessel bottom, m
D	-	Impeller diameter, m
d_p	-	Particle diameter, m
F_s	-	Volume fraction of settled solids in vessel = H_{ss}/Z
H_{ss}	-	Height of settled solids, m
N_{js}	-	Just-suspend speed, rev/s
K_{js}	-	Buurman correlating parameter defined in
L	-	Width of impeller blade, measured across the blade, m
g	-	Earth gravity, m/s²
N_{Fr}	-	Agitation Froude number = $N_{js}^2 D/g$
N_{js}	-	Just-suspend speed, rev/s
N_{jsr}	-	Relative just-suspend speed = $N_{js}/N_{js\ @\ Fs=0}$
N_p	-	Impeller power number = $P/\rho N^3 D^5$
P	-	Agitator power, watt
T	-	Vessel diameter, m
X_v	-	Volume fraction of solids in vessel
X_w	-	Weight fraction of solids in vessel
Z	-	Height of batch in the vessel, m
ε	-	Volume fraction of liquid in settled solids
ρ	-	Fluid density, kg/m³
Δρ	-	$\rho_s - \rho$, solid density - fluid density, kg/m³
HE-3	-	Three-bladed high efficiency impeller of Chemineer Inc.
MP	-	Marine propeller with a pitch to diameter ratio of 1.0
4BP	-	45 degree pitched blade impeller
6BD	-	Six-bladed disk turbine impeller

136 Industrial Mixing Fundamentals with Applications

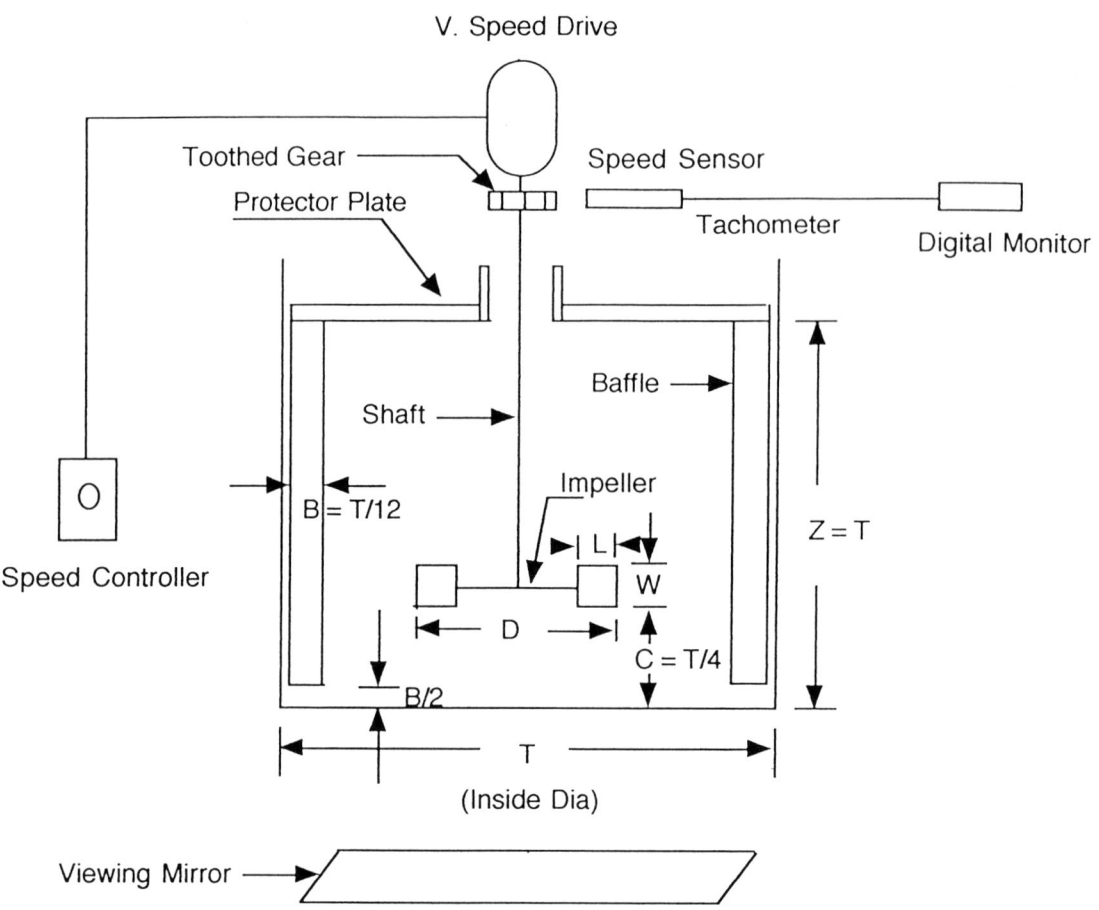

Figure 1. Schematic of Experimental Apparatus

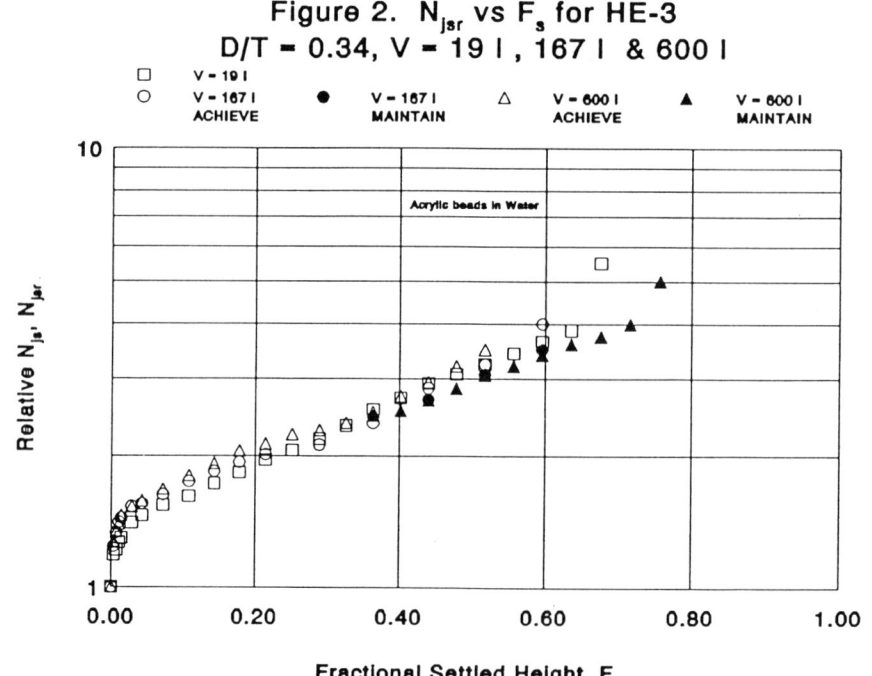

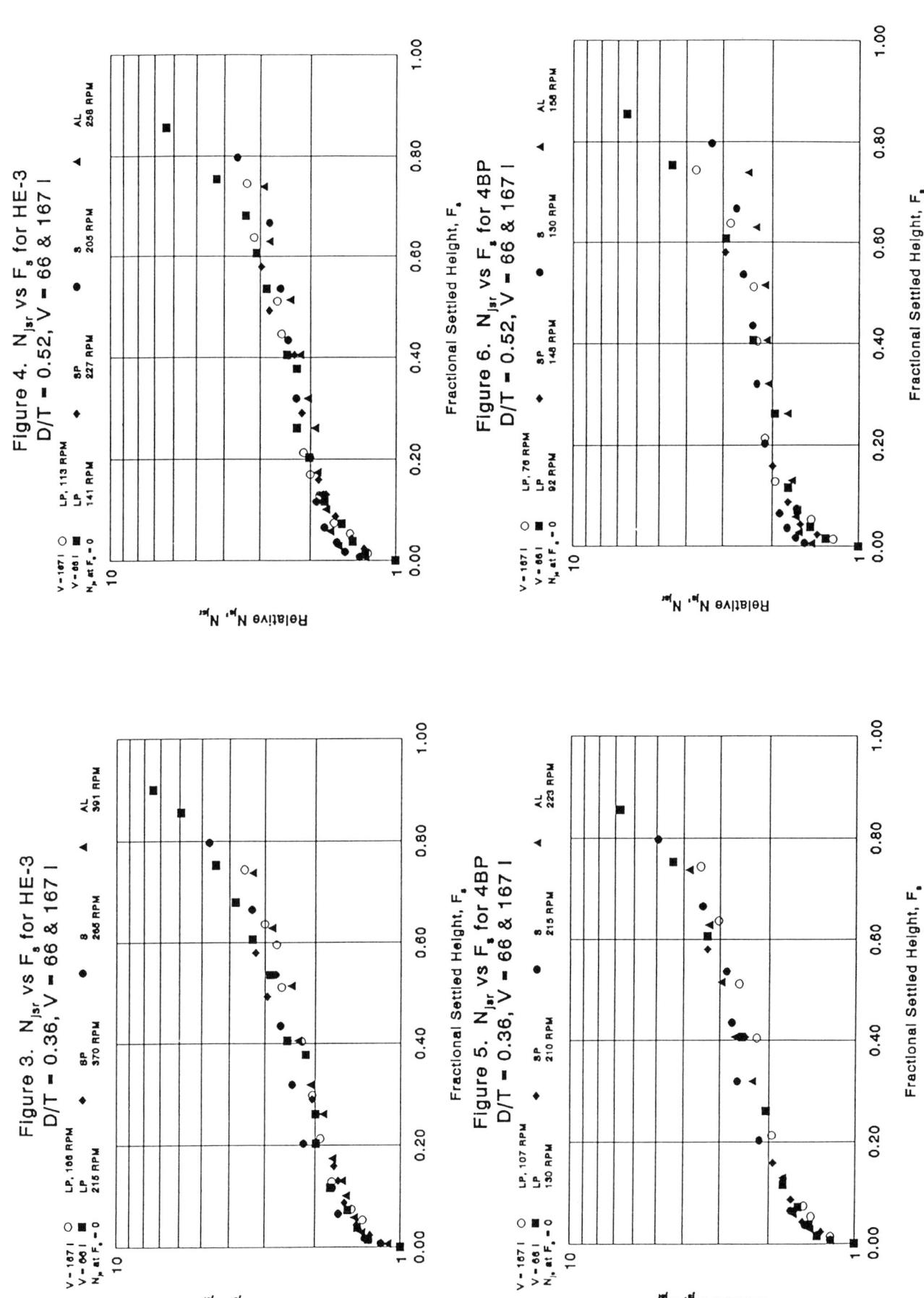

138 Industrial Mixing Fundamentals with Applications

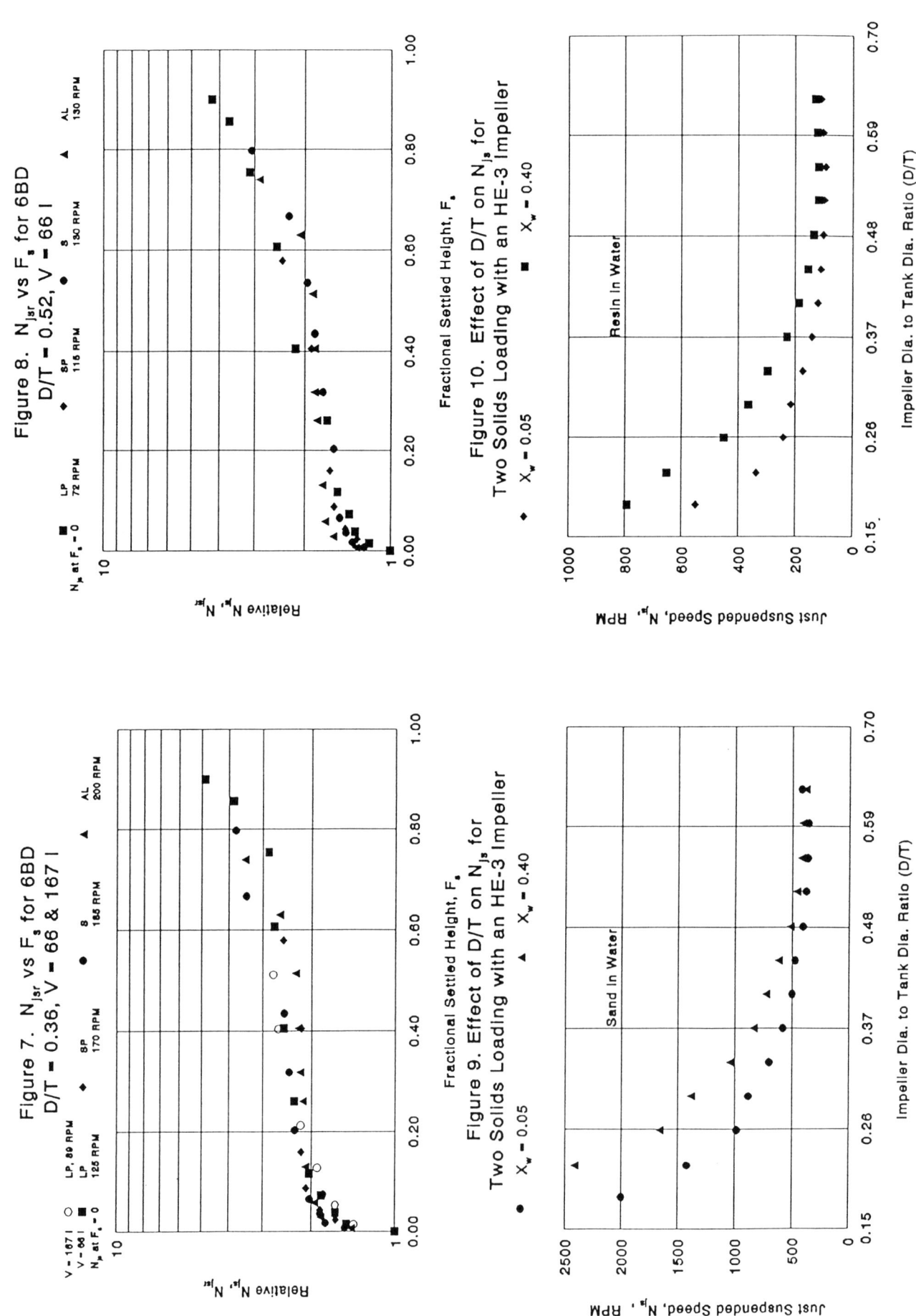

Simulation and Experimental Verification of Liquid-Solid Agitation Performance

Kevin J. Myers
University of Dayton, Dayton, OH 45469

André Bakker and Julian Fasano
Chemineer, Inc., Dayton, OH 45401

The results of the Ghost! and Fluent computational models of liquid-solid agitation have been compared with experimental solids concentration profiles obtained using a conductivity probe. Both models predict experimentally-observed local maxima and minima in the solids concentration profiles. Qualitatively and quantitatively, the three-dimensional Fluent model is the most promising, but currently its run times are excessive.

Traditionally, the design of liquid-solid agitators has relied upon laboratory testing and generalized correlations of extensive experimental data [1]. However, computational tools are now having a significant impact in this arena, and their importance is certain to increase in the future [2]. Although complex, modeling of multiphase systems has shown promise, primarily in gas-liquid systems [3]. However, to date, computational investigations of liquid-solid agitation have been limited [4]. This work investigates the performance of the Ghost! and Fluent™ models of liquid-solid agitation through comparison with experimental solids concentration profiles.

EXPERIMENTAL APPARATUS

All experiments were performed in a flat-bottomed vessel with a diameter of 0.29 meters. The solid phase was glass beads with a nominal diameter of 200 microns (ranging from 160 to 250 microns), while tap water was used as the liquid. The glass beads had a density of 2500 kg/m^3 and a measured settling velocity of 0.028 m/s. A five volume percent (11.6 weight percent) slurry was studied.

Chemineer HE-3 impellers of standard construction were used, and the impeller diameter to tank diameter ratios studied were 0.20, 0.35, and 0.57. Agitation speeds equal to 0.80, 0.90, 1.00, 1.25, and 1.50 times the just-suspended speed were considered. For one test, a four-bladed, forty-five degree pitched-blade turbine was used. Its impeller diameter to tank diameter ratio was thirty-five percent and its blade width to impeller diameter ratio was twenty percent. Square batch geometry (liquid level equal to tank diameter) was used with a fixed impeller off-bottom clearance of twenty-five percent of the vessel diameter.

Axial solid concentration profiles were measured with a two-electrode conductivity probe. The probe was positioned in the baffle plane, midway between the vessel centerline and wall (r/R = 0.5). Maxwell's equation [5] was found to describe the relation between the probe response and solids volume fraction.

COMPUTATIONAL MODELS

Early attempts to describe solids concentration profiles in agitated slurries most often relied on the sedimentation-dispersion model. Both deterministic [6] and probabilistic [7] formulations of this model have been investigated. Typically, a one-dimensional model that ignores radial and angular effects is considered. However, as computational speed has increased, the use of models that are more closely linked to the system hydrodynamics has become possible. Recently, the network of zones model [8] has been extended to liquid-solid systems.

The current generation of liquid-solid agitation models directly incorporate system hydrodynamics. The two models that are considered in this study are Ghost! and Fluent. Both two-dimensional and three-dimensional models were tested. In the two-dimensional model an axisymmetric flow pattern is assumed. In the three-dimensional model the baffles are modeled explicitly. However, this more accurate modeling comes at the cost of increased calculation times.

The Ghost! model uses Fluent to calculate a liquid-only flow pattern with impeller boundary conditions supplied from laser Doppler velocimetry. This single-phase flow pattern is then assumed to be unchanged by the presence of solids, an assumption that limits its applicability to low solids fractions. Further, particle-particle interactions are not taken into account.

Ghost! calculates the solids spatial distribution using the continuity balance for the solids. The continuity equation includes turbulent transport of solids with the turbulent diffusivity being modeled in analogy with kinetic gas theory. Rather than solving the solid momentum balance for the solid velocity field, Ghost! assumes that the solids have a constant slip velocity relative to the liquid in the axial direction, and that this slip velocity is equal to the terminal settling velocity [4].

The Fluent model is a beta version that uses a fully-coupled Eulerian two-phase approach to simultaneously determine the liquid and solid velocity and concentration fields. Continuity and momentum equations for both phases are solved simultaneously, with momentum transport between the phases being modeled with a Reynolds number-dependent, spherical-particle drag coefficient. At its current stage of development, the Fluent model requires transient simulations.

Although the Fluent model is clearly closer to first principles, its excessive run times currently limit its use to only the most critical designs. Three-dimensional simulations require two to three days on a HP-755 workstation. For this reason, the simpler, but less fundamental Ghost! model was studied in hopes that it could provide rapid simulations under some conditions.

EXPERIMENTAL RESULTS

Before comparing model simulations with experimental data, the data will be presented and examined. In the figures that present solids concentration profiles, the term dimensionless vertical position represents the distance above the base of the vessel normalized with respect to the liquid level. The average and impeller lines on the figures indicate the average solids volume percent (five) and the plane of the impeller (located one-fourth of the liquid level from the vessel base).

Figure 1 compares the solids concentration profiles of the HE-3 and pitched-blade impellers at just-suspended conditions (6.45 s^{-1} for the pitched-blade turbine and 9.78 s^{-1} for the HE-3 impeller). In this instance, the two profiles are essentially identical. It should be noted that the pitched-blade turbine requires twenty percent more power and eighty percent more torque than the HE-3 to achieve just-suspended conditions. Figure 1 also demonstrates that local maxima and minima exist in the solids concentration profiles [9].

Figure 2 illustrates the influence of the impeller diameter to tank diameter ratio on the solids concentration profile of the HE-3 impeller at just-suspended conditions (the just-suspended speeds are 28.3, 9.78, and 5.17 s^{-1} for the smallest to the largest impeller). Clearly, larger impellers produce a higher cloud height, but they also require more torque to achieve suspension [10]. Also, the local maxima and minima become more pronounced as the size of the impeller increases.

Figure 3 presents the solids concentration profiles of the smallest HE-3 impeller (D/T = 0.20) at various speeds. A significant increase in the cloud height occurs as the speed is increased from just-suspended conditions to 1.25 times the just-suspended speed. Further increase in the speed to 1.5 times the just-suspended speed does not significantly change the solids concentration profile. Also, as the speed is increased to produce a more uniform suspension, the maxima and minima in the profiles become more pronounced.

The influence of speed on the profiles of the HE-3 impeller with a diameter equal to thirty-five percent of the vessel diameter (D/T = 0.35) is shown in Figure 4. These results are similar to those of the smaller impeller except that an increase in the speed from ninety to one-hundred percent of the just-suspended speed does not cause much change in the solids concentration profile. This is most likely due to the fact that at ninety percent of the just-suspended speed only about three to five percent of the solids are not suspended.

As shown in Figure 5, the largest HE-3 impeller (D/T = 0.57) exhibits dramatically different behavior than the smaller impellers at eighty percent of the just-suspended speed. Under these conditions the solids that are suspended form a relatively uniform, dilute suspension throughout the vessel. As the speed is increased, the cloud height actually decreases as more solids are lifted into suspension. This behavior was verified visually during experimentation.

COMPARISON OF DATA AND SIMULATIONS

Figure 6 compares the solids concentration profiles predicted by Ghost! and Fluent with the experimental data obtained with the HE-3 impeller at just-suspended conditions at an impeller diameter to tank diameter ratio of thirty-five percent (D/T = 0.35). The Ghost! two-dimensional simulation shows reasonable qualitative agreement with the data, particularly relative to the cloud height. However, the three-dimensional formulation of Ghost! predicts a much more pronounced maximum concentration near the middle of the vessel and a higher cloud height, neither of which are observed experimentally.

The two-dimensional Fluent prediction does not compare well with the experimental data. In particular, this model predicts a settled solids bed over the bottom five percent of the vessel (not indicated in Figure 6). Conversely, the three-dimensional formulation of Fluent shows reasonable agreement with the experimental data.

although its concentration maximum is exaggerated and it predicts somewhat higher concentrations throughout most of the vessel, these differences are not great, and the model accurately predicts the cloud height.

The Ghost! and Fluent predictions of the complete solids concentration field in the baffle plane are presented in Figure 7, while the liquid velocity profiles of the two models are presented in Figure 8. The liquid velocity used by Ghost! is assumed to be unaffected by the presence of the solids. However, examination of the Fluent velocity profile indicates that the liquid velocity is altered by the solids. In particular, the liquid velocity above the solids cloud is dramatically reduced. Not taking this into account is most likely why the three-dimensional Ghost! model predicts cloud heights that are higher than those observed experimentally.

CONCLUSIONS

This work has yielded a number of results that impact both our knowledge of solids concentration profiles in agitated slurries and our ability to model the behavior of these systems. First, at just-suspended conditions and an intermediate impeller diameter to tank diameter ratio (D/T = 0.35), the solids concentration profiles of the pitched-blade and high-efficiency impellers are essentially identical. Also, larger impellers provide greater solids uniformity at just-suspended conditions. Further, large impellers (D/T = 0.57 for the HE-3 impeller) can yield uniform distribution of the suspended solid material throughout the vessel at levels of agitation below just-suspended conditions.

The solids concentration profile often exhibits local maxima and minima, behavior that cannot be modeled by the sedimentation-dispersion model. However, this behavior was predicted by both the Ghost! and Fluent models in their two- and three-dimensional formulations. Also, three-dimensional simulations predict higher solids suspension than do two-dimensional simulations. This was found both with Fluent and Ghost!.

Comparison of simulation predictions with experimental solids concentration profiles indicates that the quantitative accuracy of the models must be improved. Comparison of Ghost! and Fluent indicates that the influence of the solid on the liquid-phase velocity profile can probably not be ignored. Of the models tested, the three-dimensional formulation of Fluent appears to be the most promising, but its long run times make its use impractical for all but the most critical designs.

TRADEMARKS

Fluent™ is a trademark of Fluent, Inc. Lebanon, New Hampshire 03766).

ACKNOWLEDGMENTS

The authors wish to gratefully recognize the assistance of Jim Nordmeyer, Richard Ward, and Chris Wood with the experimental portion of this work.

NOTATION

D = impeller diameter, m
N_{js} = just-suspended speed, s^{-1}
r = radial position, m
R = tank radius, m
T = tank diameter, m

LITERATURE CITED

1. Myers, K. J., R. R. Corpstein, A. Bakker and J. B. Fasano, "Solids Suspension Agitator Design with Pitched-Blade and High-Efficiency Impellers", presented at the 1993 Annual AIChE Meeting, St. Louis, Missouri (1993).

2. Bakker, A., J. B. Fasano and G. M. Berg, Chem. Eng., 101, 120 (1994).

3. Bakker, A., "Hydrodynamics of Stirred Gas-Liquid Dispersions", Ph.D. Dissertation, Delft University of Technology, Delft, The Netherlands (1992).

4. Bakker, A., J. B. Fasano and K. J. Myers, "Effects of Flow Pattern on the Solids Distribution in a Stirred Tank", presented at the 8th European Conference on Mixing, Cambridge, United Kingdom (1994).

5. MacTaggart, R. S., H. A. Nasr-El-Din and J. H. Masliyah, Sep. Technol., 3, 151 (1993).

6. Barresi, A., and G. Baldi, Chem. Eng. Sci., 42, 2949 (1987).

7. Kudrna, V., V. Machoň and V. Hudcová, Coll. Czech. Chem. Commun., 45, 2070 (1980).

8. McKee, S. L., and seven others, "Measurement of Concentration Profiles and Mixing Kinetics in Stirred Tanks Using a Resistance Tomographic Technique", presented at the 8th European Conference on Mixing, Cambridge, United Kingdom (1994).

9. Godfrey, J. C. and Z. M. Zhu, "Measurement of Particle-Liquid Profiles in Agitated Tanks", presented at the 1993 Annual AIChE Meeting, St. Louis, Missouri (1993).

10. Hicks, M. T., K. J. Myers, R. R. Corpstein, A. Bakker, and J. B. Fasano, "Cloud Height, Fillet Volume, and the Effect of Multiple Impellers in Solids Suspension", presented at Mixing XIV, Santa Barbara, California (1993).

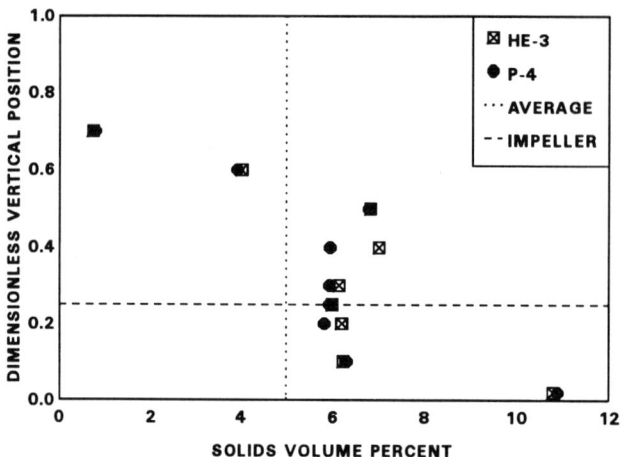

FIGURE 1: COMPARISON OF EXPERIMENTAL SOLIDS CONCENTRATION PROFILES OF HIGH-EFFICIENCY AND PITCHED-BLADE IMPELLERS AT JUST-SUSPENDED CONDITIONS FOR AN IMPELLER DIAMETER TO TANK DIAMETER RATIO OF THIRTY-FIVE PERCENT (D/T = 0.35)

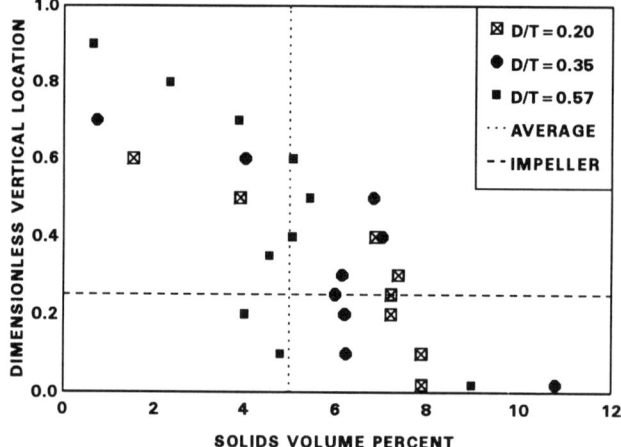

FIGURE 2: EXPERIMENTAL SOLIDS CONCENTRATION PROFILES ILLUSTRATING THE INFLUENCE OF THE IMPELLER DIAMETER TO TANK DIAMETER RATIO (D/T) AT JUST-SUSPENDED CONDITIONS FOR THE HIGH-EFFICIENCY IMPELLER

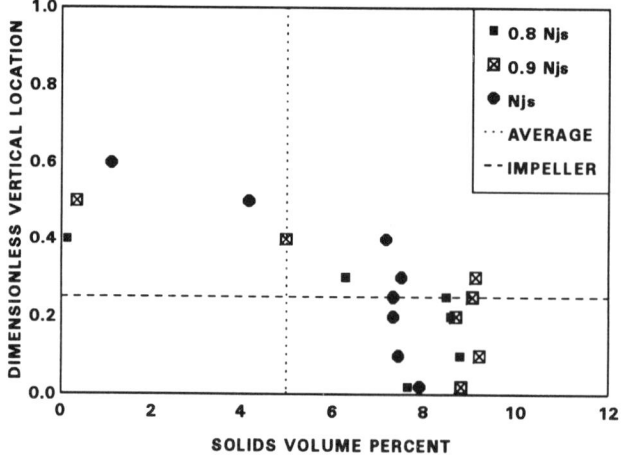

FIGURE 3A: EXPERIMENTAL SOLIDS CONCENTRATION PROFILES ILLUSTRATING THE INFLUENCE OF SPEED AT AND BELOW JUST-SUSPENDED CONDITIONS FOR THE HIGH-EFFICIENCY IMPELLER WITH D/T = 0.20

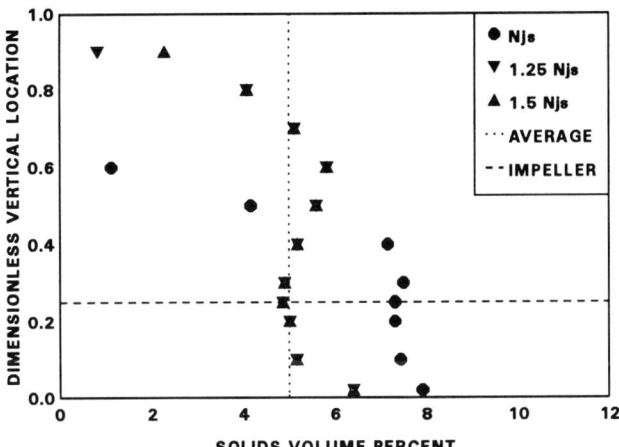

FIGURE 3B: EXPERIMENTAL SOLIDS CONCENTRATION PROFILES ILLUSTRATING THE INFLUENCE OF SPEED AT AND ABOVE JUST-SUSPENDED CONDITIONS FOR THE HIGH-EFFICIENCY IMPELLER WITH D/T = 0.20

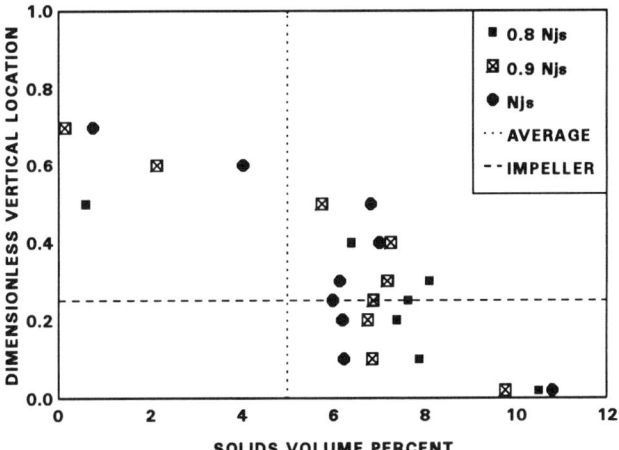

FIGURE 4A: EXPERIMENTAL SOLIDS CONCENTRATION PROFILES ILLUSTRATING THE INFLUENCE OF SPEED AT AND BELOW JUST-SUSPENDED CONDITIONS FOR THE HIGH-EFFICIENCY IMPELLER WITH D/T = 0.35

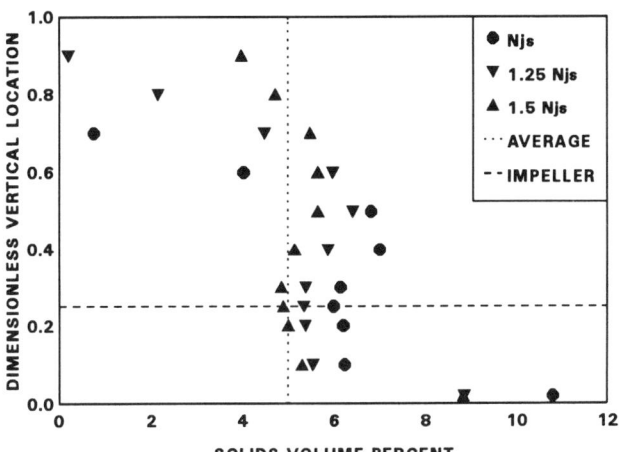

FIGURE 4B: EXPERIMENTAL SOLIDS CONCENTRATION PROFILES ILLUSTRATING THE INFLUENCE OF SPEED AT AND ABOVE JUST-SUSPENDED CONDITIONS FOR THE HIGH-EFFICIENCY IMPELLER WITH D/T = 0.35

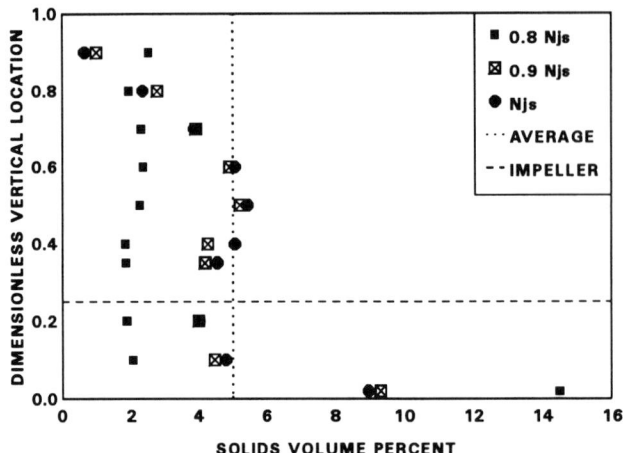

FIGURE 5A: EXPERIMENTAL SOLIDS CONCENTRATION PROFILES ILLUSTRATING THE INFLUENCE OF SPEED AT AND BELOW JUST-SUSPENDED CONDITIONS FOR THE HIGH-EFFICIENCY IMPELLER WITH D/T = 0.57

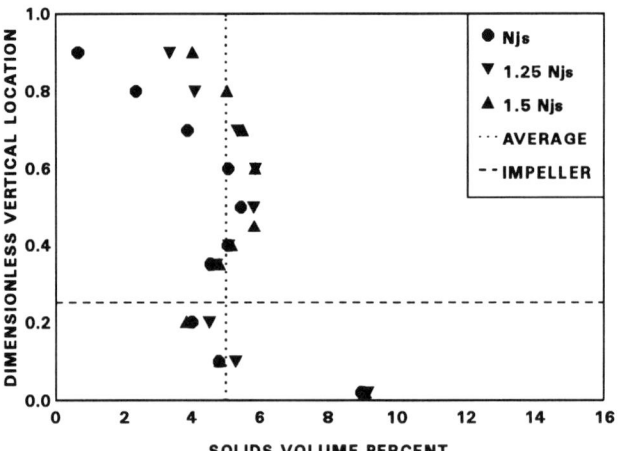

FIGURE 5B: EXPERIMENTAL SOLIDS CONCENTRATION PROFILES ILLUSTRATING THE INFLUENCE OF SPEED AT AND ABOVE JUST-SUSPENDED CONDITIONS FOR THE HIGH-EFFICIENCY IMPELLER WITH D/T = 0.57

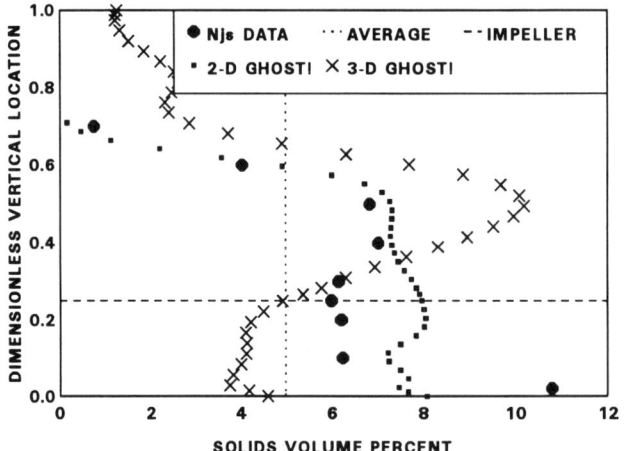

FIGURE 6A: COMPARISON OF TWO- AND THREE-DIMENSIONAL GHOST! PREDICTIONS WITH THE EXPERIMENTAL SOLIDS CONCENTRATION PROFILE OF THE HIGH-EFFICIENCY IMPELLER AT JUST-SUSPENDED CONDITIONS WITH D/T = 0.35

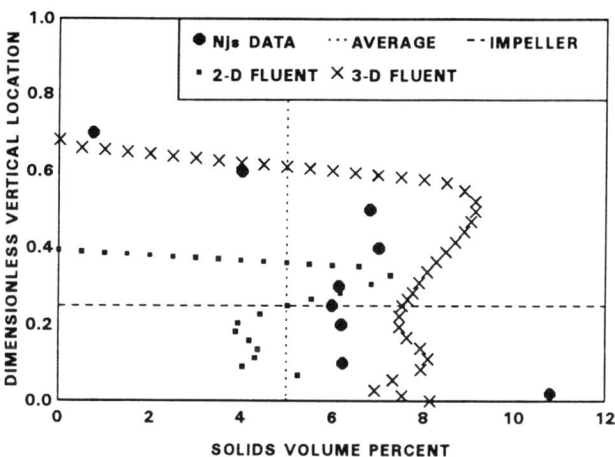

FIGURE 6B: COMPARISON OF TWO- AND THREE-DIMENSIONAL FLUENT PREDICTIONS WITH THE EXPERIMENTAL SOLIDS CONCENTRATION PROFILE OF THE HIGH-EFFICIENCY IMPELLER AT JUST-SUSPENDED CONDITIONS WITH D/T = 0.35

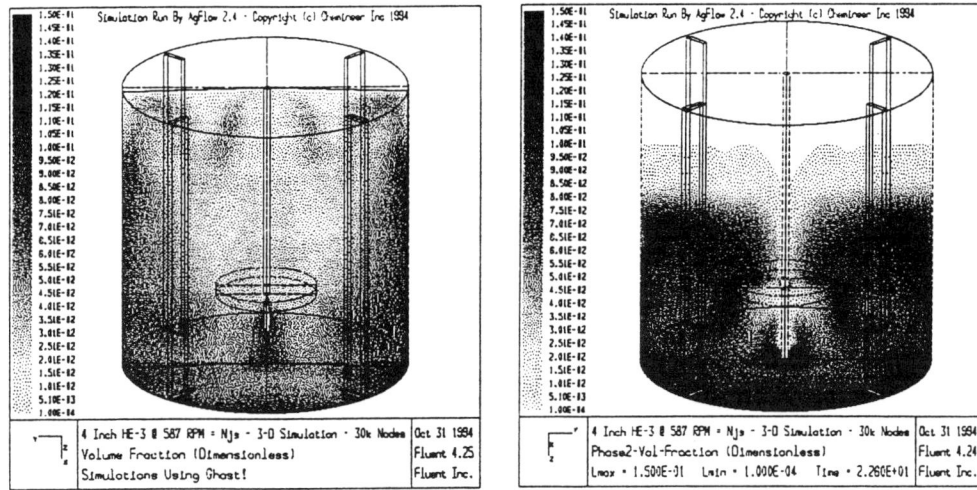

FIGURE 7: THREE-DIMENSIONAL GHOST! (LEFT) AND FLUENT (RIGHT) SOLIDS CONCENTRATION FIELD PREDICTIONS FOR THE HIGH-EFFICIENCY IMPELLER AT JUST-SUSPENDED CONDITIONS WITH D/T = 0.35

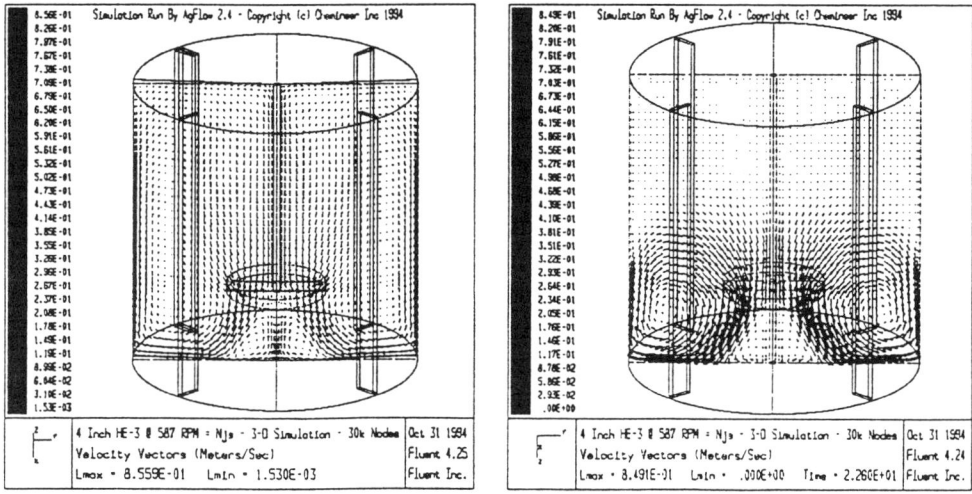

FIGURE 8: THREE-DIMENSIONAL GHOST! (LEFT) AND FLUENT (RIGHT) LIQUID VELOCITY FIELD PREDICTIONS FOR THE HIGH-EFFICIENCY IMPELLER AT JUST-SUSPENDED CONDITIONS WITH D/T = 0.35

Power Dissipation, Thrust Force and Average Shear Stress in the Mixing Tank with a Free Jet Agitator

Hanna Gladki
ITT Flygt Corporation, 35 Nutmeg Drive, Trumbull, CT 06611

The momentum transferred from a mixer jet to the liquid in the vessel produces mixing action and circulation within the tank. The jet from a submersible mixer is similar in kind as that from a nozzle and has freedom of location in the vessel. This is called Free Jet Flow Agitator (FJFA). This paper deals with the general concept of the FJFA sizing method, based on force (thrust) input from the propeller to the media.

One needs to understand the objective [3,6,9,14,17] of the fluid/solid mixing process as the first step in determining which type mixing equipment should be used. Although mixing can be accomplished by many different types of mechanisms and equipment, there are only a few types which produce effective mixing results for any individual process. We all know that there are different energy dissipation requirements for each different process. The energy dissipation characteristics for non-submersible mixers (agitators) has been covered in various books, technical-society magazines and symposium papers.

This presentation covers the use of submersible mixers (agitators) which produce a high velocity jet of liquid within a tank. The momentum transfer from the mixer jet to the liquid in the vessel causes the mixing action [6,15] and circulation within the tank. Although the submersible mixer jet is comparable to the jet from a nozzle, it holds an advantage of freedom of location in a vessel. The submersible-mixer type equipments are called Free Jet Flow Agitators (FJFA).

There has not been much information published on FJFA/mixer sizing methods. According to Tatterson [14] jet mixing is very often overlooked in mixing studies as source of mixing. Athough a mixing jet is not effective in a laminar mixing regime, approximately 70% of all flows occurring in nature are turbulent, including most processing in the chemical industry where, I believe FJFA can be used successfully.

In this presentation, I intend to discuss the general concept of the FJFA mixer sizing method, based on force (thrust) input from the propeller to the media.

The thrust is the result of force added to the system by the propeller [1,10]. The equation of motion is obtained from Newton's second law in the form which states that the sum of the external forces acting on the control volume of fluid equals the time rate of change of its momentum. (The equations used herein follow the absolute, mass-length-time, system)

$$F = m \Delta v / \Delta t$$

where: $m / \Delta t = \rho Q$.

The force exerted by the fluid in the X direction can be expressed:

$$F = \rho Q (v_{in} - v_{out}),$$

where: v_{in} = initial velocity,
v_{out} = final velocity.

If the thrust F is in the result of the action of the axial propeller on a stationary fluid volume (then $v_{out} = 0$), thrust can be expressed by the relationship:

$$F \propto \rho \, Q_j \, v_j,$$

where: ρ = density,
   Q_j = primary flow from propeller,
   v_j = jet velocity measured in the propeller blades plane.

The primary flow from the propeller is proportional to D^2 [13]:

$$Q_j \propto N D^3$$

and the jet velocity is proportional to speed N and diameter D:

$$v_j \propto N D$$

From the affinity laws the thrust force F and power P are given as:

$$F \propto \rho N^2 D^4, \qquad (1)$$

and

$$P \propto \rho N^3 D^5. \qquad (2)$$

Proportionality coefficients can be measured for each specific propeller. These coefficients are known as Power Number N_p and Thrust Number N_f.

Both the Thrust Number and Power Number are related to the Reynolds Number R_e. Although the N_p relation to R_e appears to be very well known, the N_f relation to R_e is relatively unknown. Recent studies have shown that in the turbulent zone there is very little variation of N_f with increasing R_e. In the transition zone, N_f decreases with decreasing R_e, and N_f continues an ever steeper declines with decreasing R_e in the laminar zone. Thus N_f has been found to behave similar to Flow Number N_q when the flow regime is changing from laminar to turbulent.

When Equations (1) and (2) are combined in terms of power (including proportionality coefficients N_p and N_f) they equal:

$$P = N_p / N_f \, F \, N \, D \qquad (3)$$

The force F is transmitted as a change in the velocity of the liquid mass at the tank walls and the bottom, creating pressure or so called wall shear stress τ_o [12]. An average value of the wall shear stress τ_o in the mixing tank can be expressed as equals:

$$\tau_o = F / S, \qquad (4)$$

where: S is the wall and the bottom wetted area.

When Equations (3) and (4) are combined, then specific energy ε (which equals $P/\rho V$) can be expressed as follows:

$$\varepsilon = P / \rho V = N_p / N_f (\tau_o / \rho) (S/V) \, N \, D. \qquad (5)$$

In the above equation the expression (S/V) is the inverse of the parameter known as Hydraulic Radius (R). When (S/V) in the Equation (5) is replaced by 1/R then ε equals:

$$\varepsilon = N_p / N_f (\tau_o / \rho)(N D / R) \qquad (6)$$

That R is a parameter which is related to the geometry of the tank is very well known in fluid mechanics. The larger the hydraulic radius R the more efficient is the tank geometry. Thus for the same volume of the tank the smaller wall area S will produce less resistance to flow. Therefore less energy would be needed (Equation 6) to reach the same process results. Another way of stating this is that tank geometry should be designed with the largest hydraulic radius, R, to produce the maximum flow for a given power input level.

Summarizing Equation 6, the Power Number N_p and Thrust Number N_f are functions of the type of mixer; ND relates to the speed of the mixer blade tips, and is also related to the fluid shear rate; Hydraulic Radius R describes tank geometry; the Shear Stress τ_o is the process design parameter which is a function of the process requirements and fluid properties.

DETERMINING FORCES FOR MIXING OF LOW VISCOSITY LIQUIDS.

When there are differences in composition (as density or viscosity) or temperature which create problems, mixing can be used as a mean of improving liquid uniformity in the tank/basin. The physical process which is required to achieve uniformity is called blending. The intensity of blending may be expressed by flow velocity, and simply related to scale of agitation, ranging from 1 to 10 [16]. Note that according to Dickey [3] fluid velocity in a tank is an essential parameter for quantifying mixer performance. When designing a mixing system to achieve the desired liquid uniformity, one needs to know what intensity of agitation is required, how many forces and how much power should be applied into subject tank/basin.

For determining the wall shear stress τ_o in the term of required velocity v for blending the very common equation from fluid dynamics can be used [2,18]:

$$\tau_o = \rho R (h/l) \qquad (7)$$

where l is the length of the water path in the tank, and R is the hydraulic radius.

To keep a liquid in the tank in motion an agitator creates a head, which equals the losses h in the tank. These losses h can be expressed as a function of the square of the hydraulic velocity v and a loss factor K:

$$h = K (v^2/2g) \qquad (8)$$

where the loss factor K is composed of the sum of bend losses K_b, friction losses K_f, contraction losses K_c, expansion losses K_{ex} etc. In general the value of the loss factor ($K = K_b + K_f$) in mixing tanks and basins is between 1 and 2.5, depending on their geometry.

The required level of v, which is an average hydraulic velocity, will depend on the needed intensity of agitation as derived from the 10 degrees scale identified for the different processes.

From Equations (7) and (8) we derive:

$$\tau_o = K \rho R/l (v^2/2g). \qquad (9)$$

Now, given the velocity level defined for process [16], we can use this relation for determining shear stress τ_o in the circulation channels, conduits, tanks and basins for Newtonian fluids.

Then the thrust which is required for blending can be determined from Equation (4) as follows:

$$F = \tau_o S,$$

and the specific energy ε can be defined from Equation (6).

THRUST NEEDED FOR MIXING FLUIDS POSSESSING YIELD STRESS.

The caverns formed around rotating impellers in the vessels containing yield stress fluids have been investigated by many authors [4,8]. One of the models was based on balancing the fluid centrifugal forces generated by the impeller ($\rho N^2 D^4$) with the retarding shear forces on the cavern's surface ($D_c^2 \tau_y$). Authors proposed a model for cavern size indicating that:

$$(D_c/D)^2 \propto (\rho N^2 D^2)/\tau_y \qquad (10)$$

where D_c is cavern size,

τ_y the fluid yield stress.

Etchells [5] called the dimensionless group ($\rho N^2 D^2/\tau_y$) a yield stress Reynolds number R_{ey}. The whole of the fluid in the vessel remains in motion when the cavern size reaches the vessel walls. It means that yield stress τ_y also reaches the walls.

When it is assumed that D_c^2 is proportional to the wetted surface of the mixing tank:

$$D_c^2 \propto S,$$

and existing yield stress τ_y acting on the walls equals τ_o, then Equation (10) is the same as a combination of Equations (1) and (4). This means that thrust needed to maintain the cavern volume, which approximately equals the size of the tank, is as follows:

$$F = \tau_y S \propto \rho N^2 D^4. \qquad (11)$$

An effective value for τ_y can be determined from the fluid rheology. This simple approach to define necessary thrust in the mixing tank for the fluids possessing a yield stress is applicable only for a turbulent flow in the vicinity of the propeller, where the propeller Reynolds number is 10^4, or more.

THE SUSPENSION OF SOLIDS PARTICLES

It was tested [7] for the low concentration of solids (C_w less than 10% by weight) that shear stress needed for just suspension particles off the bottom is:

$$\tau_o \propto C_w^{1/2} (\rho_s - \rho) d_p \tag{12}$$

where: C_w is the solids concentration by weight expressed in percent,

ρ_s and ρ are the density of the solids and water in kg/m^3,

d_p is diameter of a particle in m.

To define proper value for the shear stress needed for a particular process requires not only experience, but also evaluation of past studies and in many cases experimental efforts. Although defining proper shear stress value is difficult, it is the most important part of the mixer selection effort. The thrust based method produces very good results in establishing mixer applications for a given process, and when scale-up procedures from model tests are required.

ACKNOWLEDGMENT
The Author would like to acknowledge the constructive discussions with and assistance from William E. Solomon, formerly with NASA and ITT Flygt Corporation.

LITERATURE CITED

1. Albertson M.L., Barton J.R., Simons D.B., "Fluid Mechanics for Engineers", Englewood Cliffs, NJ, Prentice-Hall, 1965, pp.537.

2. Daily J.W., Harleman D.R.F., "Fluid Dynamics", Addison-Wesley Publishing Co., Inc., Reading, Mass 1966, pp.299.

3. Dickey D.S., Hemrajani R.R., "Recipes for Fluid Mixing", Chem. Eng., March, 1992, pp. 82.

4. Elson T.P., "Mixing a Fluids Possessing a Yield Stress", 6th European Conference on Mixing, Pavia, Italy, BHRA Cranfield 1988.

5. Etchells A.W., Ford W.N., Short D.G.R., "Mixing of Bingham Plastics on an Industrial Scale", Fluid Mixing III, Ins. Chem.Eng., Symposium Series, No 108 (Bradford U.K. Sept. 8-10, 1987).

6. Etchells A.W., Hemrajani R.R., Koestler D.J., Paul E.L., "The Many Faces of Mixing", Chem. Eng., March 1992, pp.92.

7. Gladki H., "Solids Suspension in the Side Entering Mixing Tank: Experimental Results", 1990 International Conference on Physical Modeling of Transport and Dispersion, MIT, Boston, 1990.

8. Harnby N., Edwards M.F., Nienow A.W., "Mixing in the Process Industries", Butterworths Series in Chem. Eng., London, 1985.

9. Koestler D.J., "Mixing and Chemical Reaction in the Production of Specialty Chemicals", Mixing Conference XIV Santa Barbara, 1993.

10. Marks' Standard Handbook for Mechanical Engineers", McGraw-Hill Book Co., 1987, pp 11-100.

11. Maruyama T., "Jet Mixing of Fluids in Vessels", Encyclopedia of Fluid Mechanics, Vol 2, N.P. Cheremisinoff(editor), Gulf Publishing Company, Houston, 1986, pp. 545-558

12. McCabe W.L., Smith J.C., Harriott P., "Unit Operations of Chemical Engineering", McGraw- Hill Book Co., NY, 1985, pp 39 and 75.

13. Oldshue J.Y., "Fluid Mixing Technology", Chem Eng., McGraw Hill Publ. Co., NY, 1983.

14. Tatterson G.B., "Scaleup and Design of Industrial Mixing Process", McGraw-Hill Publ. Co., NY, 1994.

15. Uhl, V.W., "Liquid Agitation Fundamentals", WPCF Philadephia, 23 July, 1987, Manuscript.

16. Uhl V.W., Gray J.B., "Mixing Theory and Practice", Academic Press, NY, Vol. 3, 1986, pp. 1-59.

17. Ulbrecht J.J., Patterson G.K., Mixing of Liquids by Mechanical Agitation", Gordon and Breach Science Publisher, 1985, NY.

18. White F.M., "Fluid Mechanics", McGraw-Hill Book Co.,NY, 1979, pp. 332.

The Application of CFDS-FLOW3D to Single and Multi-Phase Flows in Mixing Vessels

I.S. Hamill, I.R. Hawkins, I.P. Jones, S.M. Lo and B.A. Splawski
Computational Fluid Dynamics Services, AEA Technology, 8.19 Harwell Laboratory
Oxfordshire OX11 0RA, UK

and

K. Fontenot
Eastman Chemical Company, P. O. Box 511, Kingsport, TN 37662

Several CFD simulations of real mixing-vessel applications are presented, demonstrating the type of results that can be obtained with commercially-available CFD software. In particular the paper describes: the use of sliding and clicking meshes to model the flow around a 45° pitched-blade impeller; the validation of a two-phase model for liquid/particle flows in a draft tube baffled crystallizer; and the simulation of the three-phase flow of liquid, solid particles and sparge-gas bubbles in a multi-impeller baffled mixing vessel. All simulations were performed with the CFDS-FLOW3D software and, wherever possible, results are compared with experimental data.

Computational Fluid Dynamics has seen a considerable uptake in the Chemical and Process Industries in recent years, reflecting the growing maturity of commercial software and the demonstrable benefits gained from its application. Many of the applications, however, are for problems in relatively simple geometries and for single-phase, non-reacting flows, whilst flows in industrial mixing vessels have many complications arising from the motion of impellers and the complex physical processes which take place. This paper discusses the application of the CFDS-FLOW3D software to flows in mixing vessels and illustrates the progress which has been achieved in the modelling of these flows. This progress will be illustrated on the following problems:

The clicking-mesh and sliding-mesh modelling of a 45° pitched-blade impeller, for which experimental data are available.

Particle-laden flows in a draft tube baffled crystallizer, for which experimental data are also available.

Detailed simulation of the flow through the marine propeller used in the above-mentioned crystallizer.

The three-phase flow, of liquid, solid particles and sparge gas, in a multi-impeller mixing vessel.

MATHEMATICAL MODELS

The full Navier-Stokes equations are solved with the CFDS-FLOW3D software. The mathematical model and the solution procedure are well known and will not be repeated in detail here. The features of CFDS-FLOW3D which are of particular importance in the simulation of mixing vessel flows are the advanced turbulence models, the availability of a multi-fluid model, and the use of 'unstructured multi-block' meshes for the complex geometries of the impellers.

Impeller Modelling Options

Impellers come in a wide variety of shapes, giving rise to peculiar performance characteristics. Common to all, however, is that they impose body forces on the fluid. There are two main ways of applying these forces; (a) apply distributed momentum sources obtained from some empirical relationship; or (b) explicitly mesh the impeller geometry and rotate this section of the grid relative to the rest of the domain.

Computational Fluid Dynamics Services, Harwell Laboratory, Oxfordshire, United Kingdom

+ Eastman Chemical Company, Kingsport, Tennessee

The first method is currently easier to apply, but may be less reliable due to the empiricism in the calculation of the body force. The second method for representing the impeller can be computationally expensive, especially for multi-impeller and multiphase problems.

Explicit Modelling of Impellers

In the case of a simple cylindrical vessel without baffles, the flow does not vary with time when viewed from a reference frame rotating with the impeller. Therefore, a rotating coordinate system can be employed, with the grid fixed relative to the impeller blades and with the vessel walls moving at the same angular speed as the impellers but in the opposite direction. However, if asymmetries exist in the vessel, such as baffles or heating tubes, no reference frame exists in which the flow is stationary. Explicit modelling of the impeller and baffles must then rely on the use of a grid system which is coincident both with the fixed obstacles and the rotating impeller. Two approaches can be used to achieve this.

The first approach, which is of use only when the obstructions take the form of thin surfaces (i.e. baffles), uses a mesh which accurately represents the geometry of the impeller and which is axisymmetric and uniform in the region of the baffles. The whole mesh is considered to rotate, and the baffles can then be correctly represented by appropriately fixing the velocities at the current baffle position. The time-step size is chosen such that the grid rotates through one azimuthal cell at each time step, and the position where the baffle velocities are to be prescribed indexes accordingly. This approach is particularly attractive for laminar flows since the prescription of velocities ensures that the correct wall shear stresses are applied automatically. For turbulent flows, wall boundary conditions must also be applied. The major limitations of this approach are its restriction to simple blockage geometries and the grid-size dependency of time-step size.

The alternative approach uses two different meshes, one of which is aligned with the fixed objects, the other rotating with the impeller. At the region where these meshes meet, they slide relative to each other. Again, the meshes and time steps can be chosen such that grid points are coincident across the sliding interface, or they may be located at different locations in space, with the faces of a cell on one side of the interface being connected to more than one cell on the other side of the interface.

CFDS-FLOW3D uses multi-block data structures internally. As well as permitting the modelling of complex geometries, the multi-block approach allows great flexibility for sliding meshes, since the different mesh regions can be in different blocks, with sliding permitted along inter-block boundaries. This technique does however pose a number of numerical difficulties. The first is that it is necessary to interpolate the results across the 'sliding faces' of the mesh, though the solution of this is quite straight forward. The second is that the topology of the mesh structure changes from time step to time step, and hence the connectivity of the underlying matrix equations changes. The algebraic multi-grid method available as one of the options within CFDS-FLOW3D allows an arbitrary connectivity across grid nodes, and hence can be used to provide a robust solution technique, whatever the grid topology.

Both of the methods described above have been implemented within the software. The clicking-grid method has been coded entirely within User Fortran, whereas the sliding grids have required more extensive modifications. The sliding grid approach, however, is not as restrictive as clicking grids, and has many other applications.

Modelling of Multiphase Flows

The traditional methods for solving multiphase flows have either used a multi-fluid model, [1,2,3] with each phase having a separate velocity field and a common pressure field, or a particle tracking approach, tracking representative discrete particles or bubbles [4,5]. The multiphase examples presented in this paper employ the multi-fluid approach. The multi-fluid model solves each transport equation for each phase using the inter-phase slip algorithm of Spalding [3]. A co-located grid, with all variables located at the centre of the control volumes, is used in combination with the Rhie-Chow algorithm [6] to avoid checkerboard oscillations. Details of this approach may be found in Burns et al [7,8], and the extension to the multi-fluid approach is described by Lo [9,10], Seibert and Antal [11] and in the CFDS-FLOW3D User Manual [12]. The multi-fluid model has been widely used and validated, for example by Lo et al [13].

APPLICATIONS

Pitched-blade Turbine

The first case discussed in this paper is that of the single-phase flow of silicon oil in a cylindrical flat-bottomed baffled mixing vessel, which is filled such that the liquid depth is equal to the vessel diameter. The vessel contains four stationary baffles and a pitched-blade downward-pumping impeller. The impeller hub is attached to a coaxially-positioned shaft extending through the full depth of the vessel. The baffles, which extend throughout the liquid, are positioned symmetrically on the vessel wall, and are formed from vertically-mounted radial plates. Details of the model geometry and operating conditions are given in Table 1. The impeller Reynolds number is 20.4 and the flow regime is laminar. In this simulation, the impeller geometry is modelled explicitly using both clicking and sliding meshes, Figure 1 showing the surface mesh for the clicking-grid case.

Clicking-mesh simulation. The effect of the baffles when viewed from a fixed frame is to prevent circumferential flow and to apply radial and axial shear stresses to the fluid. In the rotating frame, zero swirl translates to a fixed angular velocity for the fluid at the baffle locations, though these locations change with time as the baffles move relative to the impeller. Therefore, to represent the baffles, their position is determined at each time step and the velocities are fixed there, the axial and radial components of the velocity being set to zero, and the swirl component to the product of the local radius and the angular speed of the rotating frame. For laminar flows such as this, prescription of the velocities at the baffle locations also has the effect of automatically creating the wall shear stresses through the built-in diffusion terms in the momentum equations. The computational grid is constructed such that the azimuthal cell size is constant, and therefore by appropriate selection of a constant time-step size the baffle positions can be made to rotate by one full cell at each time step.

The low impeller Reynolds Number of this case means that the flow from the impeller is predominantly radial and leads to the formation of toroidal vortices above and below the plane of the impeller.

The experimental data for this case are available in terms of time-averaged velocity vectors for the axial and radial velocity components, and time-averaged contours of the tangential velocity component at a vertical plane through the tank at a baffle (Lee[14]). Comparison of the experimental results and the predictions are made in Figures 2 and 3. The predicted velocity vectors of Figure 2 compare well with the measured velocities both in terms of magnitude and direction. The predicted centres of the toroidal vortices are also very close to the measured positions. The tangential velocity contours are similar to the measured values, though in the lower-velocity regions away from the impeller there is some uncertainty regarding the quality of the experimental data as evinced by the lack of smoothness of the 0.015 m/s and 0.021 m/s contours.

Sliding-mesh simulation. A sliding-mesh simulation of the same pitched-blade turbine produces results similar to those obtained with the clicking-mesh model. Figures 4 and 5 show comparisons of the results of the two approaches in terms of velocity contours at an instant when the impeller blades and baffles are in phase. Figure 4 shows contours of radial velocity in a horizontal plane through the centre of the impeller, and Figure 5 shows contours of swirl velocity in a vertical plane through one of the baffles. Slight differences in the radial velocity field may be attributed to the use of a finer grid for the sliding-mesh solution. The more significant differences in the swirl velocity at the top of the vessel arise from the use of different boundary conditions at the top surface in the two cases. In the clicking-mesh model, it is assumed that the vessel is completely filled with liquid and that the top surface is therefore a wall, whilst in the sliding-mesh simulation, a free surface is assumed and is represented as a frictionless boundary.

Crystallizer

In conjunction with its sister organisation, Separation Processes Service (SPS), CFDS has performed studies into the particle-laden flows in draft tube baffled (DTB) crystallizers. In particular, detailed comparisons have been made against experiments where spherical glass ballotini, small mono-sized particles of uniform density, have been used [15].

The cylindrical crystallizer vessel is closed by a conical base to aid particle suspension, and contains a central draft tube which is supported by radial baffles. A three-blade marine propeller is located coaxially within the draft tube, and generates upflow or downflow depending on the direction of rotation. Figure 6 shows a sketch of the crystallizer, together with the location of the measurement points. The multi-fluid model is used to represent the solid and liquid phases, and the momentum-source method is employed for the representation of the impeller effects. Figure 7 shows the predicted flow pattern for the liquid phase in three different vertical planes, and indicates that there is little azimuthal variation of the velocity field around the crystallizer. The flow pattern for the solid phase is similar to that of the liquid phase, though with slightly higher downflow velocities and lower upflow velocities because of the settling.

Figures 8 and 9 present comparisons of the predicted and measured particle concentrations for low and high mass loadings of particles respectively. The agreement is good at low loadings but is less satisfactory at higher loadings, when experiments indicate greater sedimentation occurring than in the predictions. The discrepancy may be due to uncertainties in the representation of the propeller, or to the particle drag correlation which neglects particle-to-particle interactions and volume fraction effects. Further work is in hand to test more sophisticated models for the interphase drag and to calibrate precisely the effect of the propeller upon the flow.

Crystallizer Propeller

Two-phase simulations of the SPS crystallizer experiments indicate that the distribution of the solid phase is sensitive to the details of the momentum-source representation of the propeller. To obtain better understanding of the operation of the propeller, a three-dimensional explicit model has been created of the propeller, draft tube and crystallizer base, and the results of this simulation will be used for calibration of the momentum sources in the full two-phase crystallizer model. Data describing the propeller geometry was obtained from a three-axis digital inspection system, and was read into the SOPHIA grid generator as surface files, from which the computational grid was then constructed. The computational solution domain does not extend to the baffled region of the crystallizer, and the stationary walls are therefore axisymmetric. Hence, a coordinate system rotating with the propeller can be used, avoiding the need for the more expensive sliding-grid models.

Provisional results have been obtained (Spares [16]), and Figure 10 shows the geometry of the propeller and velocity vectors on a vertical plane through the draft tube. In Figure 11, radial profiles of axial velocity are presented at azimuthal intervals of 30° in plane A of Figure 10. At this plane, which is approximately one chord-length below the propeller, both radial and circumferential variation in the axial velocity is apparent. The general trend however is for the axial velocity to increase with radius. Figure 12 shows similar profiles in plane B of Figure 10, at which point the circumferential variation in axial velocity is negligible. The near axisymmetry of the flow at the bottom of the draft tube indicates that a circumferentially-averaged, but radially varying, momentum source should be able to correctly represent the effect of the propeller on the flow in the bulk of the vessel.

Slurry Reactor

Work has taken place recently to model the three-phase flow of solids, liquid and gas in a mixing vessel agitated by five impellers. The complexity of the system precludes the use of explicit modelling of the impellers on the grounds of computational expense, and therefore the momentum-source approach is adopted. The stirred tank is comprised of a cylindrical vessel with an ellipsoidal base, four vertical baffles, and a coaxially-mounted shaft carrying one radial and four axial impellers. A slurry composed of a liquid laden with small particles is injected on the side wall of the vessel, while a toroidal sparge ring injects gas. The model employs the standard multiphase features of CFDS-FLOW3D to simulate the three-phase flow (solids, liquid and gas bubbles), and turbulence is represented by the standard two-equation k-ε turbulence model.

A multiblock mesh is used to model the geometry of the vessel accurately, including internal baffles and the sparge ring. The impeller momentum sources are calculated from the power input to the shaft, and are distributed amongst the different impellers according to the mean volume fractions in each of the impeller swept volumes. Locally, mass-fraction weighting of the momentum sources amongst the phases ensures that the same specific impulse is applied to each of the fluids.

The high absolute pressure in the vessel relative to the variation in the hydrostatic pressure means that the dependence of bubble size upon the total head is negligible. However, the bubbles are broken up as they pass through the impellers, and this effect is represented by the solution of a scalar variable in the gas phase representing the mean bubble size.

The main features of the flow are evident in the liquid-phase velocity vector plot of Figure 13. The vortices at the tips of the four axial impellers are clearly visible, as is the flow induced by the rising bubbles. The gas-phase volume fractions, Figure 14, reveal that the bubbles initially move upwards through the tips of the bottom radial impeller, and are then entrained by the first axial impeller. However, their buoyancy prevents them from following the liquid flow down into the bottom impeller, and consequently a core of low gas concentration forms in this region. The particle volume fractions (normalized by the inlet particle concentration in Figure 15) show a significant amount of settling occurring.

These simulations enable the effect of different features, such as impeller type and position and the location of baffles and sparge rings, to be investigated, and the model to be used as a tool for improving the design.

CONCLUSIONS

A selection of CFD simulations of mixing-vessel applications has been presented, including single- and multi-phase flows, and implicit and explicit modelling of impellers.

Explicit sliding-grid simulation of impellers can be computationally expensive, but provides detail of the flow in these regions which is unavailable with momentum-source models, and can yield such useful information as local turbulence levels and shear rates. In applications such as droplet breakup or the mixing of easily-damaged cells, this type of information may be critical to the accurate prediction of vessel performance. From the point of view of the impeller manufacturer, explicit simulations of new impeller designs can complement experimental tests and reduce development times and costs.

When large-scale mixing information is required, the details of the flow in the impellers may be less important, and momentum-source models provide a practical and less expensive modelling approach. However, the accuracy of such models depends on the characterization of the impeller behaviour, and here the sliding-grid approach can be employed to determine the momentum flux and turbulence profiles generated by the impeller.

The complexity of many multiphase flows compared to their single-phase equivalents, and the difficulties which are often encountered in obtaining experimental data in real multiphase processes, make them ideal subjects for CFD analysis. The multi-fluid model has been proven and validated on a large range of multiphase problems, and here is shown to give good agreement with crystallizer experiments at low solids loadings. However, the poorer agreement at higher loadings emphasizes the importance of accurately representing the effect of the impeller.

LITERATURE CITED

1. Harlow F.H. and Amsden A.A., "Numerical calculation of multi-phase fluid flow", J. Comp. Phys **17** pp 19-52, 1975.

2. Harlow F.H. and Amsden A.A., "Flow of interpenetrating material phases", J. Comp. Phys. **18** pp 440-464, 1975.

3. Spalding D.B., "Two phase momentum, heat and mass transfer in chemical process and energy engineering", ed F. Durst, Hemisphere Press, Washington, 1979.

4. Livesley D.M., Oakley D.E., Gillespie R.F., Elhaus B., Ranpuria C.K., Taylor T., Wood W. and Yeoman M.L., "Development and validation of a computational model for spray-gas mixing in spray dryers", Proc. Drying '92 Conference, pp 407-415, 1992.

5. Benim A.C. and Neuhoff H.G., "Analysis of erosion behaviour in a turbocharger radial turbine", Int. J. Num. Meth. in Fluids, **16** pp 259-285 1993.

6. Rhie C.M. and Chow W.L., "Numerical study of the turbulent flow past an airfoil with trailing edge separation", AIAA J. **21** pp 1527-1532, 1983.

7. Burns A.D. and Wilkes N.S., "A finite difference method for the computation of fluid flows in complex three-dimensional geometries", AEA Technical Report AERE-R 12342, 1987.

8. Burns A.D., Wilkes N.S., Jones I.P. and Kightley J.R., "FLOW3D: body-fitted coordinates", AEA Technical Report, AERE-R 12262, 1986.

9. Lo S.M., "Mathematical basis of a multi-phase flow model", AEA Technical Report AERE-R 13432, 1989.

10. Lo S.M., "Multiphase flow model in the Harwell-FLOW3D computer code", AEA Technical Report, AEA-InTec-0062, 1990.

11. Siebert B.W. and Antal S.P, "An IPSA-based two-fluid algorithm for boiling multi-phase flows", presented at First CFDS User Conference, 1993.

12. CFDS-FLOW3D User Manual, Available from Computational Fluid Dynamics Services.

13. Lo S.M, Hannan M., Hallas N., Jones I.P. and Wilkes N.S., "Multi-phase CFD applications in the process industry", Proceedings WUA-CFD Conference, Basel, 1994.

14. Lee, C., Private Communication, 1994.

15. SPS Symposium, Separation Processes Service, Harwell Laboratory, 1992.

16. Spares R., "Explicit methods for the modelling of fluid flows around impellers", MSc. Thesis, University of Teesside, 1994.

Table 1: Geometry and operating conditions for the pitched-blade impeller mixing vessel.

Notation	
vessel internal diameter [cm]	14.5
fluid depth [cm]	14.5
No of baffles	4
baffle width [cm]	1.25
baffle thickness [cm]	0.3
fluid density [kgm^{-3}]	1000
fluid dynamic viscosity [$kgm^{-1}s^{-1}$]	0.211
impeller speed [rpm]	100
No of blades	4
blade external diameter [cm]	5.08
blade pitch angle [°]	45
blade width [cm]	0.9
blade thickness [cm]	0.1
distance from bottom impeller edge to vessel floor [cm]	6.67
hub diameter [cm]	1.68
hub height [cm]	1.1
shaft diameter [cm]	0.8

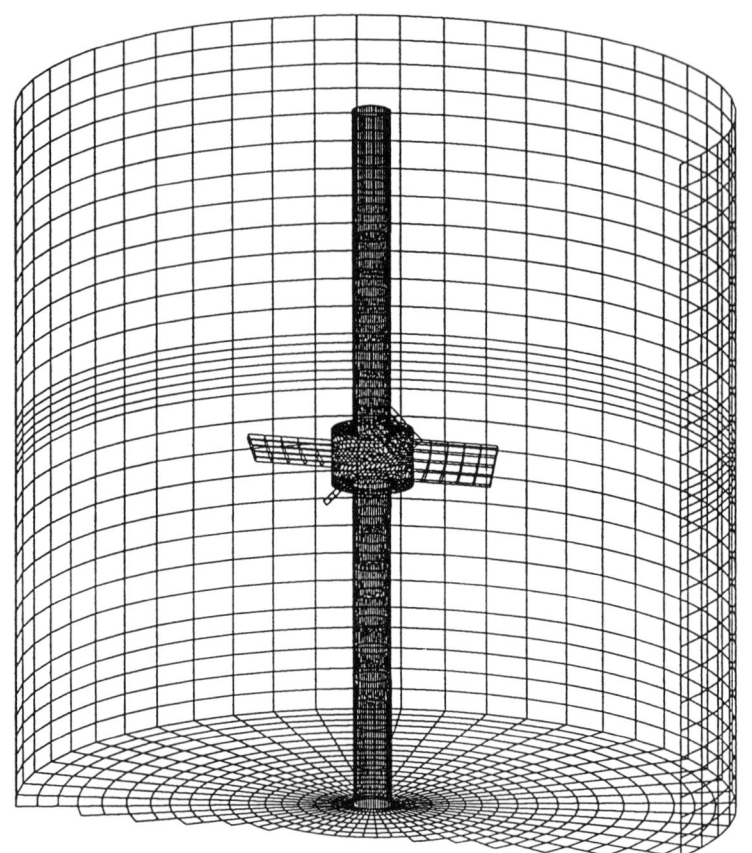

Figure 1. Surface mesh for clicking-grid model of pitched-blade turbine.

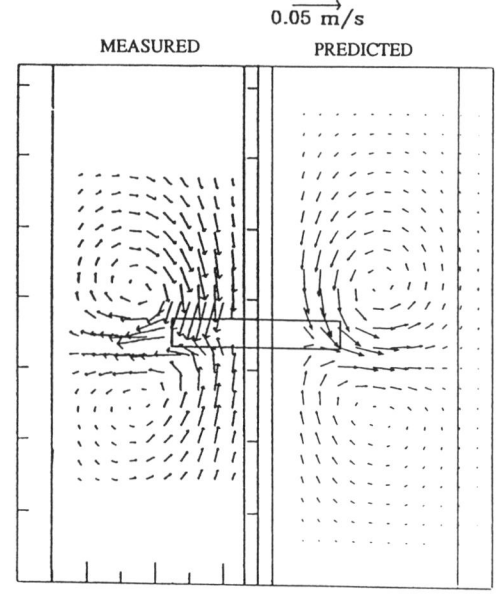

Figure 2. Measured and predicted time-averaged velocity vectors.

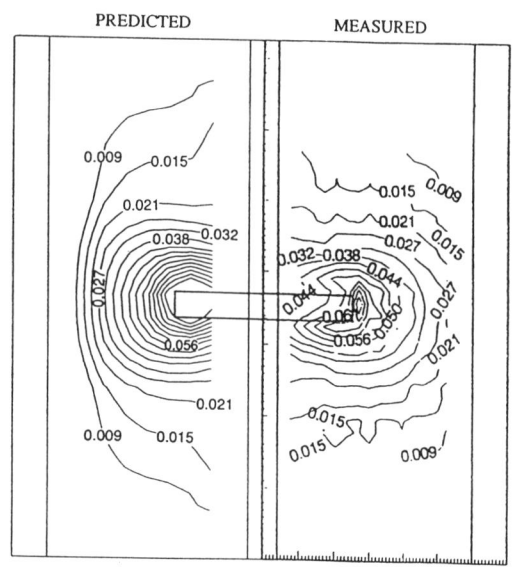

Figure 3. Measured and predicted time-averaged contours of swirl velocity.

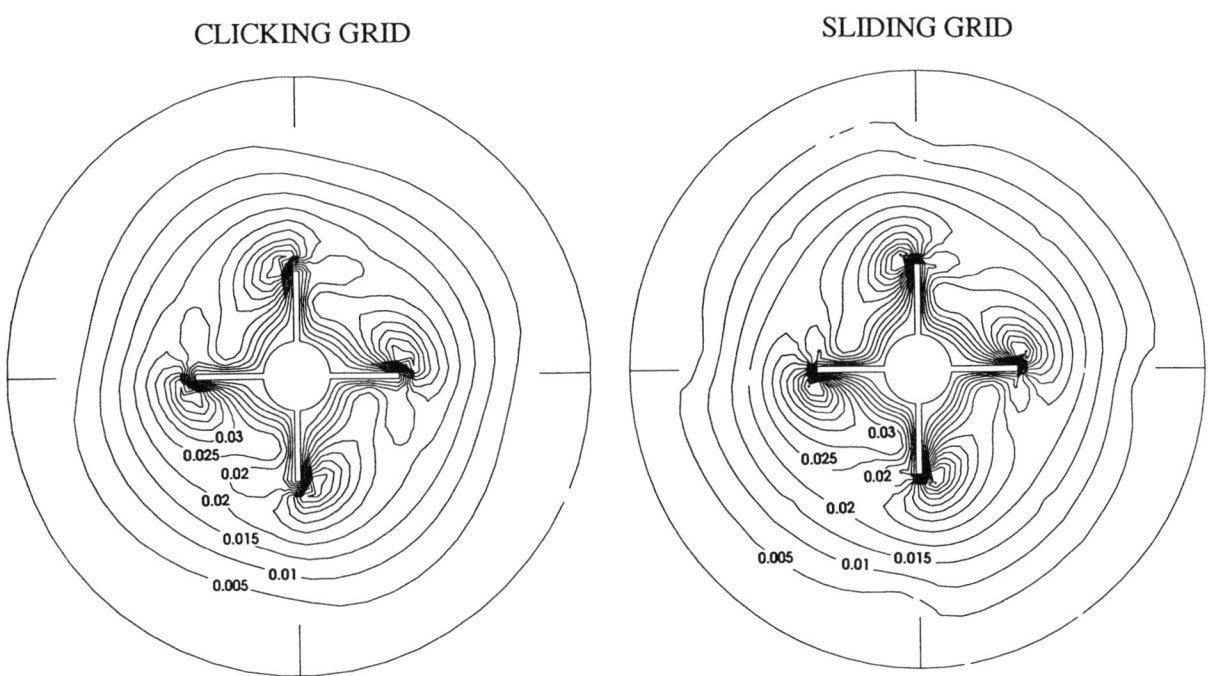

Figure 4. Comparison of clicking-grid and sliding-grid results - contours of radial velocity.

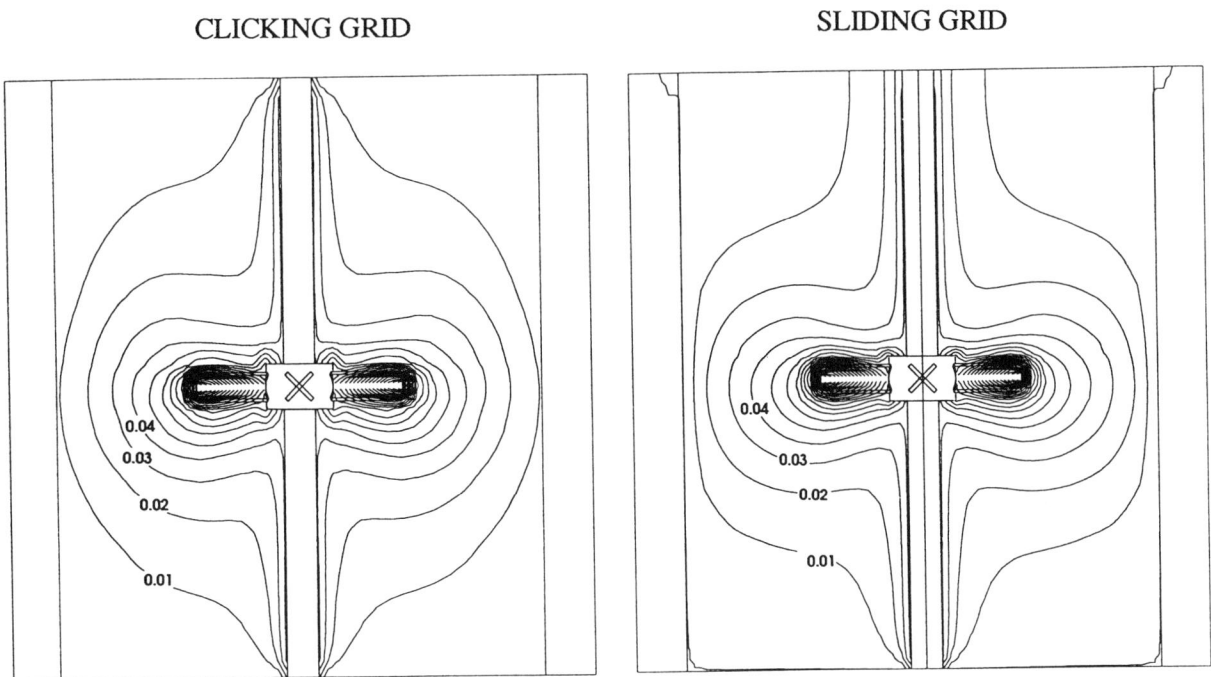

Figure 5. Comparison of clicking-grid and sliding-grid results - contours of swirl velocity.

158   Industrial Mixing Fundamentals with Applications

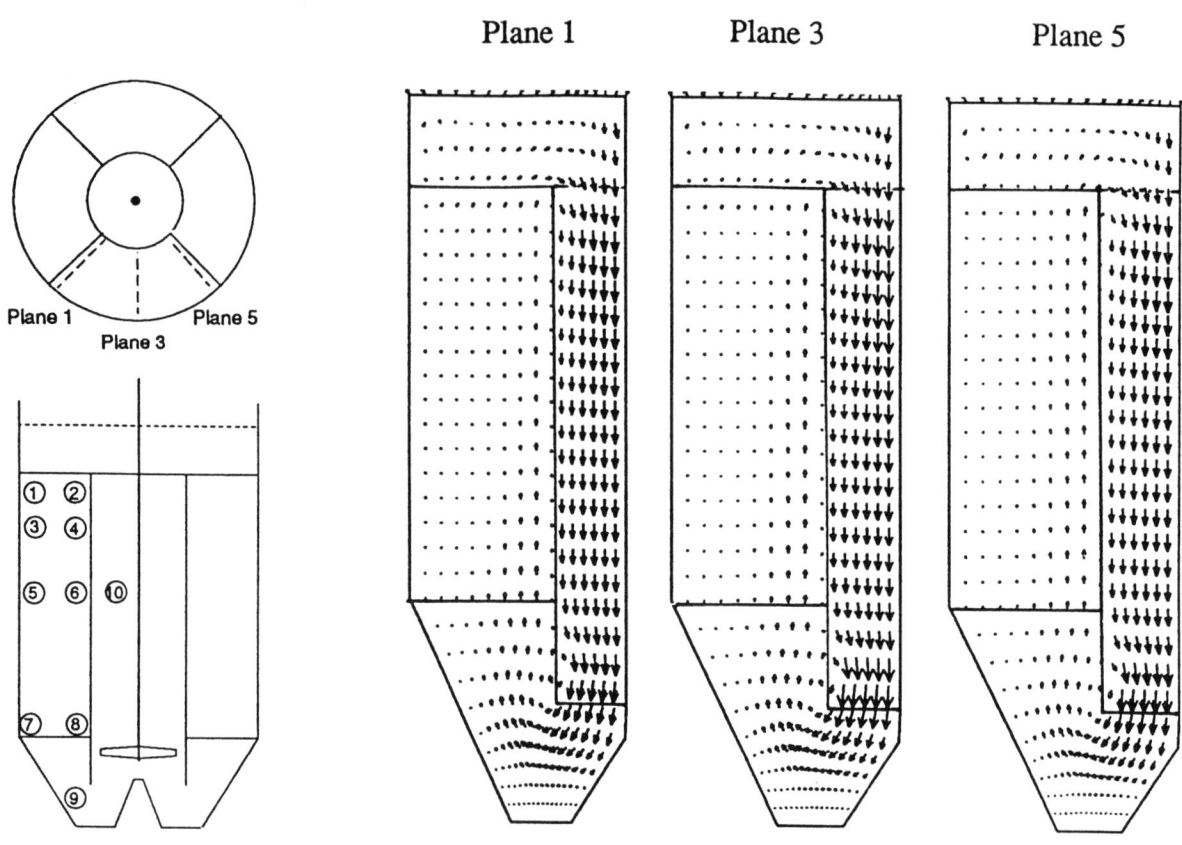

Figure 6. Crystallizer geometry.

Figure 7. Liquid-phase velocities at three vertical planes.

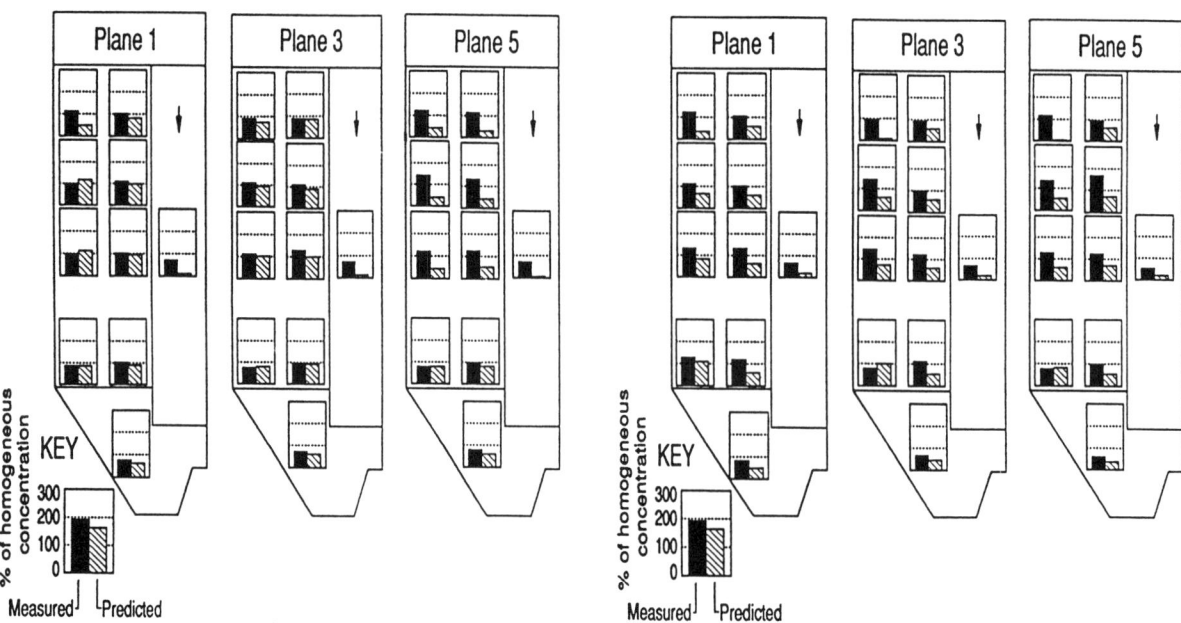

Figure 8. Measured and predicted particle volume fractions- 0.5 volume %, 130μm ballotini at 200rpm.

Figure 9. Measured and predicted particle volume fractions- 3.0 volume %, 430μm ballotini at 500rpm.

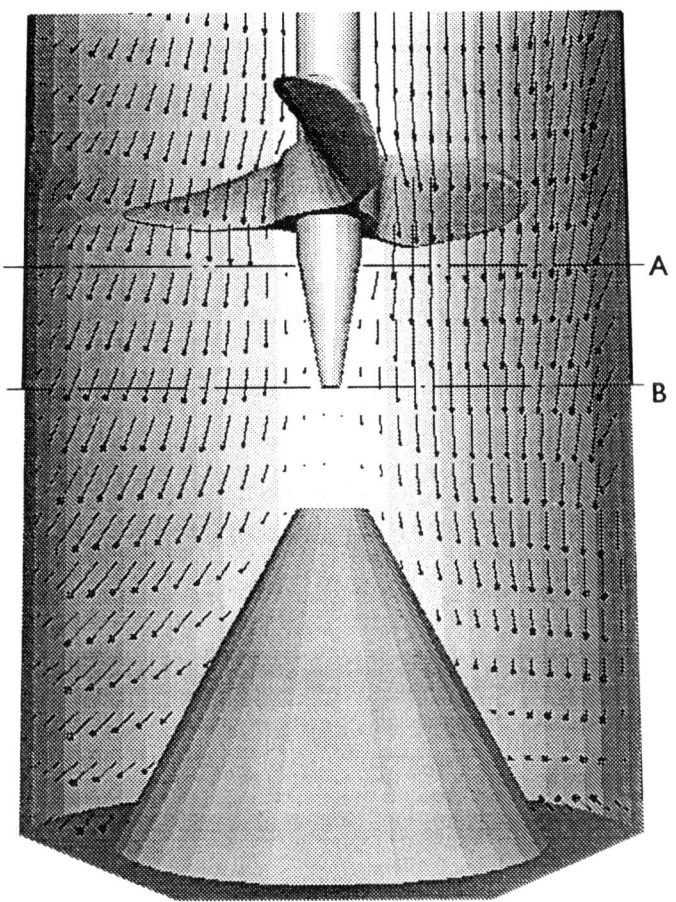

Figure 10. Geometry of crystallizer propeller and velocity vectors on vertical plane.

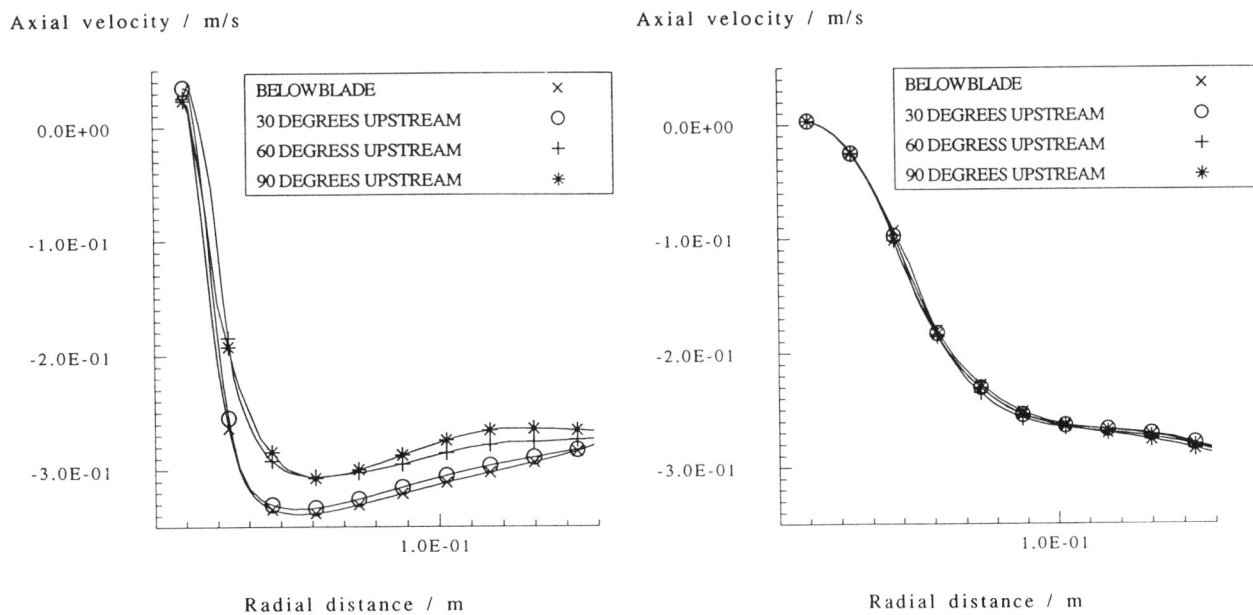

Figure 11. Radial profiles of axial velocity at plane A.

Figure 12. Radial profiles of axial velocity at plane B.

160 Industrial Mixing Fundamentals with Applications   AIChE SYMPOSIUM SERIES

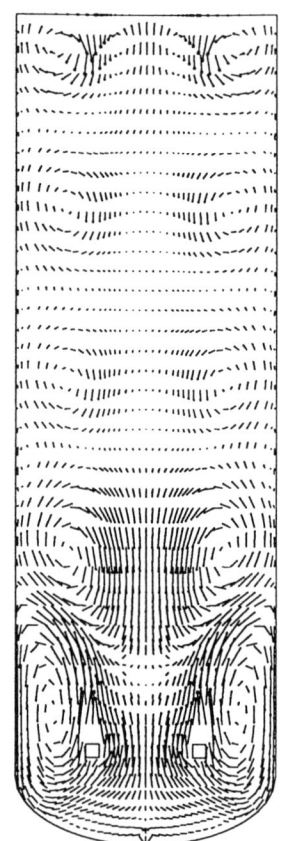 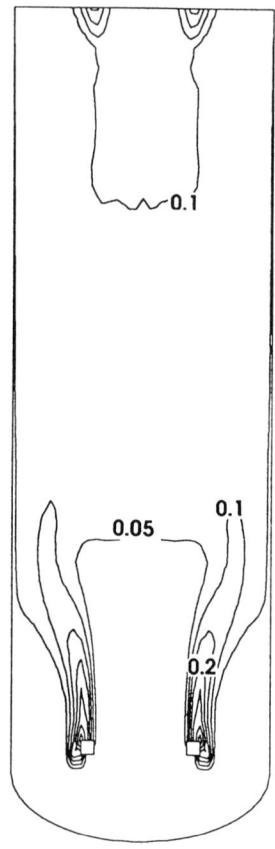 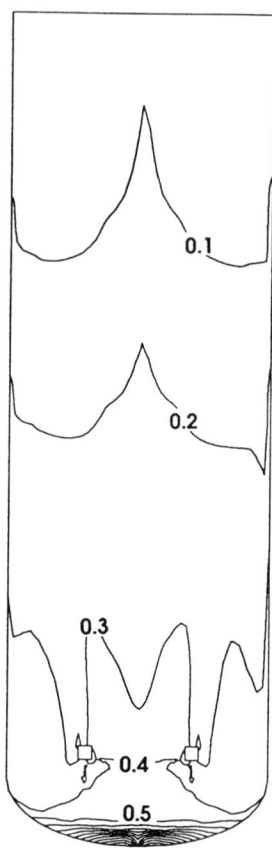

Figure 13. Liquid-phase velocity vectors.

Figure 14. Volume fraction of bubbles.

Figure 15. Normalized particle volume fractions.

Gas-Liquid Mixing and Mass Transfer in Tall Tanks

C. Kurt Svihla, R. Eric Berson and Thomas R. Hanley
Speed Scientific School, University of Louisville, Louisville, KY 40292

Gas phase tracer tests cannot be used to determine residence time distributions in agitated gas-liquid systems unless the results are corrected to account for mass transfer of the tracer gas. Distortion of the apparent residence time distribution occurs even for relatively insoluble gases and becomes more pronounced as the ratio of the liquid volume to the gas flow rate increases. This paper presents the results of experiments conducted in a sparged, multi-turbine agitated vessel with an aspect ratio of 3:1. Gas hold-ups were measured experimentally and compared with apparent values determined from the moments of gas phase tracer tests. The apparent residence time distributions were corrected for the effects of mass transfer using an approximate analytical approach. Tests were conducted in tap water and for two different concentrations of cellulose fibers selected to mimic the rheological behavior of mycelial broths. The feasibility of using tracer tests to determine gas phase hold-ups and residence time distributions is examined.

This study involved the characterization of the gas phase hold-up and apparent residence time distributions for tap water and cellulose fiber suspensions in a sparged, agitated vessel (0.298 m diameter) with a liquid height to tank diameter ratio of 3:1. Hold-ups were evaluated manometrically and from analysis of photographs showing the position of the dispersion surface relative to fixed scales at several points within the vessel. Gas phase residence time distributions were measured by using pulse tests with a helium tracer. An attempt was made to correct the apparent residence time distribution for the effect of mass transfer by fitting the apparent residence time distributions to a model which accounted for finite rates of mass transfer. The $k_L a$ values needed for this analysis (for helium) were estimated from the values measured for oxygen using a dynamic absorption technique

EXPERIMENTAL

Tests were conducted in a sparged, baffled, Plexiglas vessel agitated by three Rushton turbines. The tank diameter, T, was 0.298 m and the liquid volume was 0.0626 m³ with a 3:1 aspect ratio (H/T = 3). Four baffles with diameters equal to T/10 were present. The diameters of the turbines were all equal to D = T/3. The lowest impeller in the flat-bottomed vessel was located at an off-bottom clearance of T/3; the remaining two turbines were spaced at intervals of T. Gas was introduced through a six-port ring sparger with a diameter equivalent to that of the turbine disks.

Agitator rotation rate and power input were measured by means of a rotating torque sensor equipped with optical torque and speed transducers (Vibrac, Inc). Gas flow rates were measured by electronic mass flow meters (Unit Instruments). Helium concentrations in the gas phase tracer tests were measured using a thermal conductivity detector (Gow-Mac, Inc.). All electronic signals were interfaced to a Metrabyte DAS-16 interface board. The switching of valves in the gas phase tracer tests and $k_L a$ measurements was controlled by the computer using relays, solenoid valves, and pneumatically actuated valves.

Tests were conducted in tap water and for 10 kg/m³ and 20 kg/m³ suspensions of cellulose fibers (Solka-Floc, grade KS-1016, James River Corporation). The manufacturer quotes an average length of 290 μm for this particular fiber grade. This choice of fiber type was based on a recommendation in the literature (Chisti and Moo-Young, [1]) that suspensions of Solka-Floc fibers could adequately represent the rheological behavior of *Aspergillus niger* broths.

University of Louisville, Louisville, Kentucky, Eric Berson is now at Vanderbilt University, Nashville, Tennessee.

Methods

Gas phase residence time distributions were determined by recording the response of the gas leaving the dispersion surface to a pulse of helium tracer as measured by a thermal conductivity detector. The exit gas was sampled through an inverted sampling funnel placed at the dispersion surface. The observed response included the effects of the sampling and detection system as well as the RTD of the dispersed gas phase. A separate pulse was introduced to the sampling funnel and the resulting normalized response deconvoluted from the overall response using FFT analysis (least squares fitting of the convolution of the sampling and detection system response with an assumed model to the actual overall system response).

Gas phase hold-ups were evaluated manometrically and from analysis of photographs showing the position of the dispersion surface relative to fixed scales at several points within the vessel. As many as nine points were used for the visual evaluation of hold-up with the result reported as a linear average of the individual readings.

RESIDENCE TIME DISTRIBUTIONS

The experimental normalized impulse response curves were fit to a transfer function consisting of three first-order terms (well-mixed regions) with a dead time, or plug-flow region:

$$E(s) = \frac{e^{-t_d s}}{(\tau_1 s + 1)(\tau_2 s + 1)(\tau_3 s + 1)} \quad (1)$$

The response of this system to a normalized impulse response is zero for $t < t_d$ (dead time). For $t > t_d$:

$$E(t) = \frac{\tau_1}{(\tau_1 - \tau_2)(\tau_1 - \tau_3)} e^{-(t-t_d)/\tau_1} + \frac{\tau_2}{(\tau_2 - \tau_1)(\tau_2 - \tau_3)} e^{-(t-t_d)/\tau_2} + \frac{\tau_3}{(\tau_3 - \tau_1)(\tau_3 - \tau_2)} e^{-(t-t_d)/\tau_3} \quad (2)$$

This response has mean and variance given by:

$$\tau_E = \tau_1 + \tau_2 + \tau_3 + t_d \quad (3)$$
$$\sigma_E^2 = \tau_1^2 + \tau_2^2 + \tau_3^2 \quad (4)$$

If there is some finite solubility of the tracer gas in the liquid, the true residence time distribution can not be obtained directly from a tracer test. Mass transfer of the tracer gas will distort the apparent RTD by reducing the peak height and broadening the tail. The apparent mean residence time determined from the first moment of the tracer response curve will be higher than the value given by the ratio of the gas phase hold-up volume divided by the average volumetric flow rate. Nocentini and co-workers [2,4] suggested that the increase in mean time is independent of the value of the mass transfer coefficient:

$$\mu_1 = \left[1 + \frac{\alpha(1-\varepsilon)}{\varepsilon}\right] t_R \quad (5)$$

where α, the tracer solubility constant, is defined in terms of the Henry's Law constant as $(RTC_{solvent})/\mathcal{H}$. Eqn (5) may be rearranged to yield:

$$\mu_1 = \left[t_R + \frac{\alpha V_L}{Q_G}\right] \quad (6)$$

This expression will hold only for a concentration detector with infinite resolution, however. It must fail, for example, as $k_L a \to 0$. The magnitude of the actual *observed* effect of mass transfer upon the apparent residence time distribution will therefore be a function of both $k_L a$ and the resolution of the concentration detector.

In order to simulate the effect of mass transfer upon the apparent RTD in more detail, a model was developed by including mass transfer terms for each region of the assumed RTD and assuming a well-mixed liquid phase. The resulting equations may be written:

$$\frac{\partial C_{G1}}{\partial t} + \frac{q_G}{\varepsilon A} \frac{\partial C_{G1}}{\partial z} = \frac{k_L a_1}{V_{G1}} (\alpha C_{G1} - C_L) V_L \quad (7)$$

$$\frac{dC_{G2}}{dt} = \frac{q_G}{V_{G2}} (C_{G1} - C_{G2}) - \frac{k_L a_2}{V_{G2}} (\alpha C_{G2} - C_L) V_L \quad (8)$$

$$\frac{dC_{G3}}{dt} = \frac{q_G}{V_{G3}} (C_{G2} - C_{G3}) - \frac{k_L a_3}{V_{G3}} (\alpha C_{G3} - C_L) V_L \quad (9)$$

$$\frac{dC_{G4}}{dt} = \frac{q_G}{V_{G4}} (C_{G3} - C_{G4}) - \frac{k_L a_4}{V_{G4}} (\alpha C_{G4} - C_L) V_L \quad (10)$$

$$\frac{dC_L}{dt} = \int_0^L \frac{k_L a_1}{L} (\alpha C_{G1} - C_L) dz + \sum_{i=2}^{i=4} k_L a_i (\alpha C_{Gi} - C_L) \quad (11)$$

Here region 1 is the plug flow region while regions 2, 3, and 4 are the three-well-mixed regions.

The equations for an impulse input can be solved either numerically, or, if desired, by an approximate analytical approach. An approximate analytical solution can be obtained using Laplace transforms if all the dead time terms which appear in the denominator of the expression for the Laplace transform of the outlet gas concentration are replaced by rational polynomial (i.e. Padé) approximations. Problems which are stiff numerically typically require higher order Padé approximations for good accuracy for times close to the plug-flow residence time. The rather lengthy manipulations required to obtain a solution in the Laplace domain and subsequently invert it into the time domain were performed using MAPLE V (Release 3).

The model results were fit to the apparent curves primarily by adjusting two of the three well-mixed region time constants (the apparent plug-flow region volume was assumed to be unaffected by mass transfer). The $k_L a$ values for helium were estimated by multiplying the values measured for oxygen using a variant of the dynamic method described elsewhere (Svihla and Hanley, [4]) by an appropriate function of the ratio of the diffusivities of helium and oxygen. The multiplying factor was allowed to vary between $\sqrt{D_{He}/D_{O2}}$ and D_{He}/D_{O2}, in accordance with the bounds set by penetration theory and film theory, respectively.

RESULTS AND DISCUSSION

Figure 1 shows the comparison of the hold-up values determined by the manometric technique and the visual determination of the level rise upon aeration for tap water at 11 and 19 slm. Values calculated from a recent correlation reported by Pinelli et al. [5] for a similar system are included for comparison. The two techniques tend to agree at higher impeller speeds, with the visually determined values providing a better match to the literature correlation over the whole range of speeds. Figure 2 shows the effect of the fiber concentration on the experimentally measured gas-hold-ups. There is little apparent decrease in the observed hold-up for the 10 kg/m^3 suspensions, but the hold-up values do appear to decrease for the 20 g/L suspensions, especially at higher impeller speeds. This is in broad agreement with the results of Chisti and Moo-Young [1] who studied the effect of cellulose fiber concentration on the observed gas phase hold-up in bubble columns and air-lift reactors.

Figure 3 shows the space time calculated from the measured gas phase volume fraction and the apparent mean times from the gas phase tracer tests corrected according to eqn (6). At low speeds (and hence low values of $k_L a$), the corrected mean times from the apparent RTD are generally lower than the mean times calculated from the gas flow rate and hold-up, while at higher speeds (and higher $k_L a$ values), the two mean times tend to be in closer agreement. Since the magnitude of the *observed* effect of mass transfer on the apparent residence time distribution is known to increase with $k_L a$, this behavior is consistent with expectations.

Table 1 shows the comparison of the hold-up measurements with those determined by correcting the apparent residence time distributions for the effect of mass transfer by solving the set of partial differential equations given as eqns (7) - (11). It is interesting to note that while the results are in reasonable agreement in many instances (with the values at 10 g/L being a notable exception), there is a systematic bias in that the values from the tracer tests are always greater than the corresponding values from the hold-up measurements. This may be due to the model assumption of a well-mixed liquid phase (the mixing times in the vessel appear to be on the order of a few to several seconds which is not insignificant in terms of the time-scales of a tracer response test), or may be caused by some other factor which has heretofore been neglected in the analysis (surface aeration or non-uniform surface cross-section). Until the reason for this bias can be determined, it seems premature to conclude that the actual gas phase residence times have been determined by this approach and analysis.

CONCLUSIONS

Gas phase hold-ups for gassing rates of 11 and 19 slm and agitation speeds from 220 to 520 rpm do not appear to change as the fiber concentration is increased from 0 to 10 kg/m^3 but do appear to decrease as the fiber concentration increases further to 20 kg/m^3. The model for mass transfer developed in this paper could not satisfactorily explain all the differences between the mean times of the apparent residence time distributions and the mean times based on the measured hold-ups and gas flow rates.

ACKNOWLEDGMENT

This work was partially supported by a grant from the National Science Foundation (# BCS-9196147).

NOTATION

A	tank cross-sectional area, m^2
C	concentration, $kmol/m^3$
D	impeller diameter, m
D_{He}	diffusivity of helium in water, m^2/s
D_{O2}	diffusivity of oxygen in water, m^2/s
E	normalized impulse response, s^{-1}
H	ungassed liquid height, m
$\mathcal{H}$	Henry's Law constant
$k_L a$	volumetric gas-liquid mass transfer coefficient, s^{-1}
Q	volumetric flow rate, m^3/s
t	time, s
t_d	dead time or plug-flow residence time, s
t_R	true mean gas phase residence time, s
T	tank diameter, m
V	volume, m^3
z	axial coordinate in tank, m

Greek Symbols

α	dimensionless tracer solubility constant
ε	fractional gas phase hold-up
μ_1	mean time from impulse response curve, s
τ_i	time constant of ith well-mixed region, s

Subscripts

G	gas phase
L	liquid phase

LITERATURE CITED

1. Chisti, M.Y., and M. Moo-Young, *Biotechnol. Bioeng.*, **31**, 487 (1988).

2. Nocentini, M., G. Pasquali, and F. Magelli, *Colloque Agitation Mecanique*, 5-77 (1986).

3. Nocentini, M., F. Magelli, and G. Pasquali, *Sixth Euro. Conf. Mixing*, (BHRA, 337, 1988).

4. Svihla, C.K., and T.R. Hanley, *AIChE Symp. Ser.*, **88** (#286), 114 (1992).

5. Pinelli, D., M. Nocentini, and F. Magelli, *IChemE Symp. Ser.*, #136, 81 (1994).

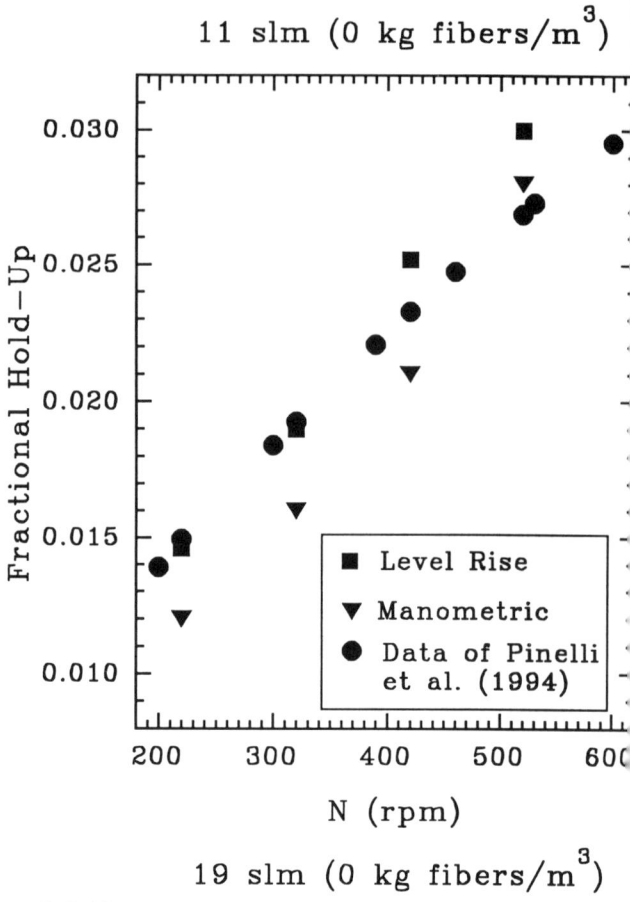

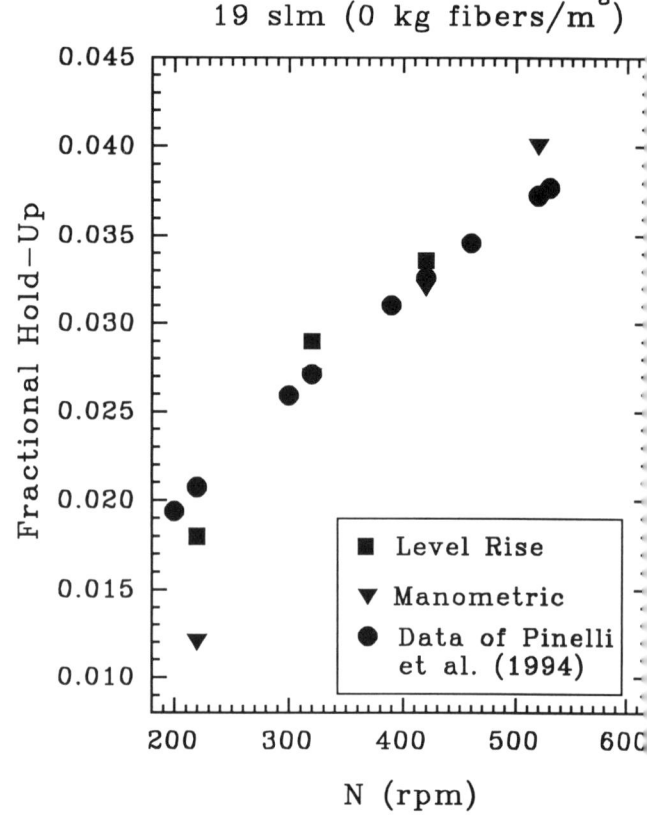

Figure 1. Comparison of the results of different hold-up measurement techniques with data reported the literature.

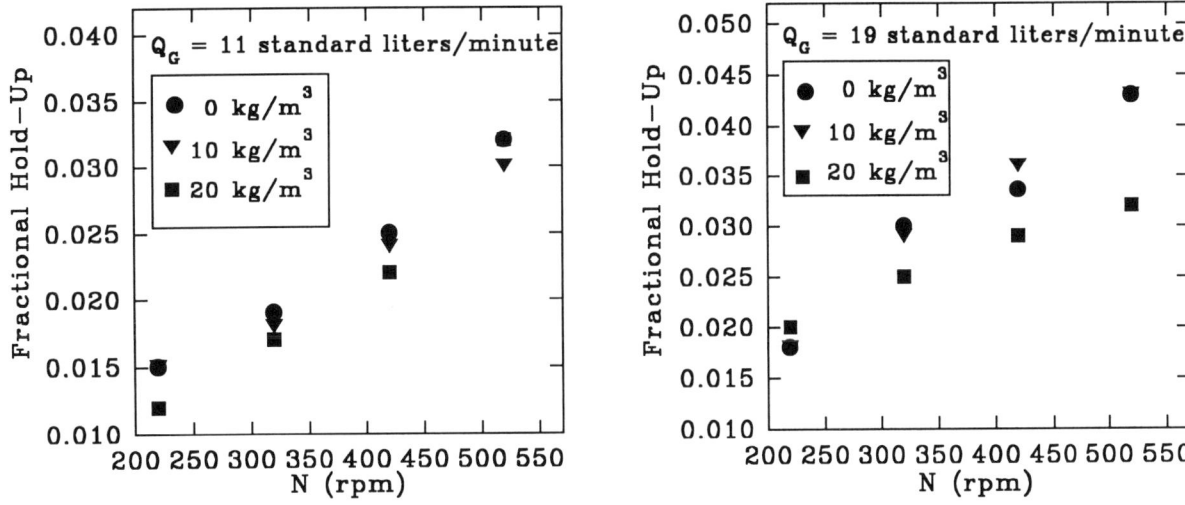

Figure 2. The effect of cellulose fiber concentration upon the measured fractional gas phase hold-up at 11 and 19 slm

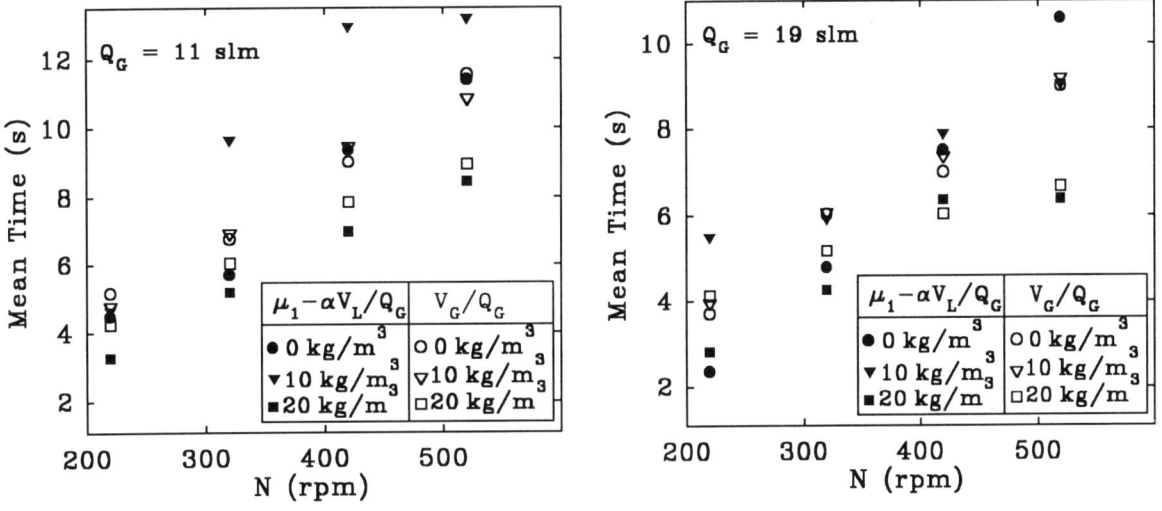

Figure 3. Mean times determined from the measured hold-ups and the tracer response curves corrected using eqn (6)

Table 1. Comparison of the measured hold-ups with those determined from the corrected residence time distributions

C_{fiber} (kg/m³)	Q_G (slm)	N (rpm)	Measured Hold-Up	Hold-Up from RTD	$\frac{(k_L a)_{He}}{(k_L a)_{O_2}}$	C_{fiber} (kg/m³)	Q_G (slm)	N (rpm)	Measured Hold-Up	Hold-Up from RTD	$\frac{(k_L a)_{He}}{(k_L a)_{O_2}}$
0	11	220	0.015	0.021	1.85	10	19	220	0.018	0.035	1.84
0	11	320	0.019	0.024	1.83	10	19	320	0.029	0.036	1.70
0	11	420	0.025	0.033	1.84	10	19	420	0.036	0.044	1.68
0	11	520	0.032	0.038	1.89	10	19	520	0.043	0.049	1.60
0	19	220	0.018	0.021	1.89	20	11	220	0.012	0.018	3.67
0	19	320	0.030	0.033	1.77	20	11	320	0.017	0.023	3.60
0	19	420	0.034	0.042	1.89	20	11	420	0.022	0.027	3.60
0	19	520	0.043	0.051	1.84	20	11	520	0.032	0.030	3.66
10	11	220	0.015	0.022	1.84	20	19	220	0.020	0.022	3.75
10	11	320	0.018	0.033	1.84	20	19	320	0.025	0.029	1.88
10	11	420	0.024	0.042	1.83	20	19	420	0.029	0.037	3.60
10	11	520	0.030	0.043	1.87	20	19	520	0.032	0.038	2.14

Hydrodynamic Shear Stress Effects on the Actin Cytoskeleton and Energy Status of Cultured Epithelial Cells

Vinayak D. Bhat, Robert S. Cherry
Center for Biochemical Engineering, Duke University, Durham, NC 27708

Lazaro J. Mandel
Department of Cell Biology, Duke University Medical Center, Durham, NC 27708

In bulk bioreactors, mammalian cell cultures are agitated to enhance oxygenation and mixing which are essential for optimal growth. The turbulence that results from the agitation, however, exposes the cells to significant shear stress. Steady unidirectional shear forces have been seen to alter the stress fiber architecture and orientation of endothelial cells. Because of their size and being derived from a lower shear environment, the epithelial cells may be more sensitive to the effects of shear stress. The present project examined the effect of hydrodynamic shear on the actin cytoskeleton structure and the energy content in cultured epithelial cells. A spinner flask system was used to expose Madin-Darby canine kidney (MDCK) cells to fluctuating shear stress, which was measured using hot-wire anemometry. The impeller Reynolds number of 700 indicated that a near turbulent environment was experienced by these cells. Confocal microscopic and biochemical analysis demonstrated that the F-actin structure collapsed and cells detached after 9-12 h of shear stress exposure. Since metabolic depletion of cellular ATP effects the F-actin network, alteration of the metabolic state of the cells was investigated as an intermediary step for the shear stress effects. Stress fibers became thinner and eventually disappeared after 60 minutes of ATP depletion and lead to a loss of epithelial attachment. Despite this correlation, the cellular ATP content was not depleted in cells exposed to shear stress. This study demonstrates that external hydrodynamic forces reorganize the cytoskeleton of MDCK cells by a mechanism independent of ATP depletion.

The biotechnological potential of animal cell-derived products such as hormones, interferons, monoclonal antibodies, virus vaccines, and recombinant-DNA proteins has driven the development of bulk bioreactor production of mammalian cell cultures. Currently approved drugs derived from bulk mammalian cell bioreactors include erythropoietin, tPA, OKT3, and Factor VIII. Enhancing cell viability and efficiency within these bioreactors remains the focus of industrial and scientific efforts.

Mixing in bioreactors increases mass transfer of oxygen from gas phase to the liquid phase and provides a homogeneous environment for cell growth. Due to the absence of a cell wall and their relatively larger size than microbial cells, mammalian cells are more sensitive to hydrodynamic forces.

R.S. Cherry is now with Idaho National Engineering Labs, Idaho Falls, ID.

Correspondence concerning this paper should be addressed to L.J. Mandel.

Although fluid dynamics have been intensively studied for standard chemical engineering applications, they have not been applied to mammalian cell bioreactors. Unlike chemical systems, biological systems have a highly nonlinear response to excessive fluid forces which can result in cell death. Understanding the effect of the hydrodynamic shear forces on the cell's physiology and structural integrity is critical for improving bioreactor design and scale-up.

Midler and Finn [1] were the first to report the effects of hydrodynamic forces on cells in viscometers and reactors for protozoa cells. Since then, there have been many reports on the effect of well-defined flows on freely suspended (cells which can grow in suspension) and anchorage dependent cells (cells which require a substrate for growth). Laminar shear stresses affect the cell shape, cytoskeleton structure and physiology of mammalian cells (Levesque and Nerem [2], Stathopoulos and Hellums [3]). The confluent monolayers of porcine aortic endothelial cells elongated in the direction of shear at shear stress of 20 dyne/cm², the rate and amount of elongation increased proportionately with increasing shear stress (Ookawa et al. [4]).

The cytoskeleton is a complex network of protein filaments that extend throughout the cytoplasm of a cell. This internal cytoskeletal framework of the cell provides structural supports for the forces applied to the cell, conducts intracellular transport, allows the cell to alter its shape and enables cellular movement in a coordinated and directed fashion. Principally, the cytoskeleton is made up of microtubules, intermediate filaments and microfilaments. Each type of filament is formed from a different protein subunit: tubulin for microtubules, a family of fibrous proteins for intermediate filaments and actin for microfilaments.

The microfilaments consist of a tight helix of uniformly oriented, globular subunits called G-actin (MW = 43,000 Da) which is associated with a molecule of non-covalently bound ATP. When present in sufficient concentration, non-covalently bound to ATP and in the presence of monovalent and divalent cations, primarily K⁺ and Mg²⁺, G-actin polymerizes into filamentous actin or F-actin [Alberts et al. [5]. In the basal layer of cultured cells, bundles of actin filaments form the stress fibers. The stress fibers insert themselves at one end into the plasma membrane at special junctions called focal contacts and the other end into a meshwork of intermediate filaments that are localized in perinuclear region or into a second focal contact. At the site of a focal contact, the cell is attached to the extracellular matrix through transmembrane linker proteins of the integrin family. These linkages, which are indirect and through various attachment proteins, are the physical connections between the cytoplasmic domain and the external domains of the cell (Figure 1). A primary consequence of the microfilaments in anchorage-dependent cells is the ability to withstand mechanical stresses. For example, treatment of hybridoma cells with cytochalasin E or B, high ammonia or low pH disrupts their F-actin network and renders them much more sensitive and vulnerable to shear stress (Peterson [6]).

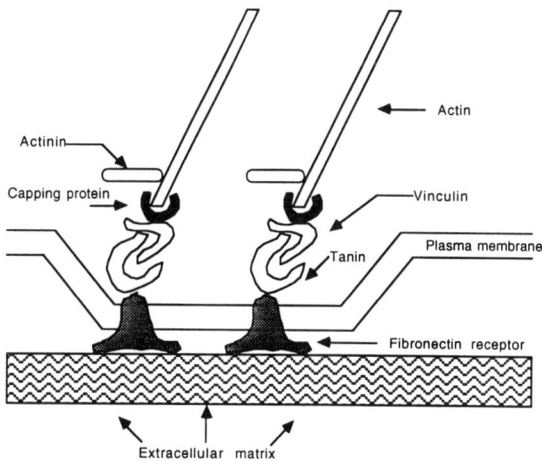

Figure 1: A model of transmembrane linker glycoproteins in the plasma membrane connecting intracellular actin filaments to the extracellular matrix at a focal contact.

A great deal of variation seems to exist between different cell types and specific reactor/impeller configurations (Cherry and Papoutsakis [7]). In cultured endothelial cells, which are derived from cells lining the blood vessels and experiencing shear stress *in vivo*, shear stress causes significant effects (Dewey et al. [8], Eskin et al. [9]). Far less is known about the effects of shear stress on cultured epithelial cells, which are derived from cells lining the inner and outer surface of the body and which see minimal shear forces *in vivo*. Madin-Darby canine kidney (MDCK) cells, a distal tubular epithelial cell line with a well-characterized three dimensional cytoskeletal structure was utilized for an epithelial model (Bacallao et al. [10]). A primary objective of these studies was to determine the effects of shear stress on the F-actin network and the degree of cell detachment in MDCK epithelial cells.

These studies determined that the F-actin network of the basal stress fibers and the strength of attachment of MDCK cells were very sensitive to low levels of shear stress. The cellular mechanisms that underlied the stress fiber depolymerization and cell detachment remained undetermined. A similar degree of basal F-actin depolymerization and loss of cell attachment was reported by Bacallao et al. [11] when MDCK cells were depleted of ATP. The secondary objective of these studies was to determine if shear stress initiated a depletion in the cellular ATP content which would instigate the F-actin depolymerization and cell detachment. Interestingly, shear stress did not alter the cellular ATP content suggesting that the effects of shear stress acted via a pathway that was independent of ATP depletion.

METHODS AND MATERIALS

General Experimental Design:

The cells were seeded on glass cover slips and transferred to the spinner flask after they reached confluency. The cells were removed at different time intervals, fixed, stained and observed under confocal microscopy and compared with the controls which were placed in the no-stress environment. The F-actin content in the cells at different time intervals of shear stresses was quantified by comparing the fluorescence intensity. Constant temperature hot-wire anemometry was used to measure the shear stress experienced by the cells inside the spinner flask. The ATP and protein content in the cells were measured at the different time intervals and compared with the controls.

Cell culture techniques

Madin Darby canine kidney (MDCK) cells were cultured in 200 ml sterile culture flasks at 37 °C and 5% CO_2 in air until they became confluent. Dulbecco's Minimal Essential Medium [DMEM] supplemented with 5% (v/v) new born calf serum, 1% (v/v) penicillin and 1% (v/v) streptomycin sulfate was used as growth medium. When confluency in the flask was achieved, the cells were removed using trypsin-EDTA and seeded on glass cover slips in six-well plates. The medium used for growing these cells on the cover slips was the same as above with an additional 1 µl/ml serum extender Mito [+] (Collaboration Research Inc.). The cover slips with confluent cells were placed in the spinner flask system. One cover slip from each batch was kept outside in a no-stress environment and served as the control. The cover slips were removed from the spinner flasks at different time intervals and were stained for F-actin or assayed for ATP content.

Spinner Flask Apparatus

The 150 ml spinner flask system (Corning) used in this work produces near-turbulence with low mean shear stress. The spinner flask has two annular Teflon rings sitting at the bottom. The external ring is fixed and supports the internal ring. The slots provided on the internal ring hold the glass cover slips (22 x 22 mm) in a vertical position (Figure 2). The MDCK cell monolayers were placed inside the spinner flask facing the magnetic impeller. All the experiments were carried out at an impeller speed of 20 rpm.

Fixation and Staining

The cells are fixed with 0.5 % glutaraldehyde in a solution of 80 mM K-PIPES, pH 6.8, 5mM EGTA,

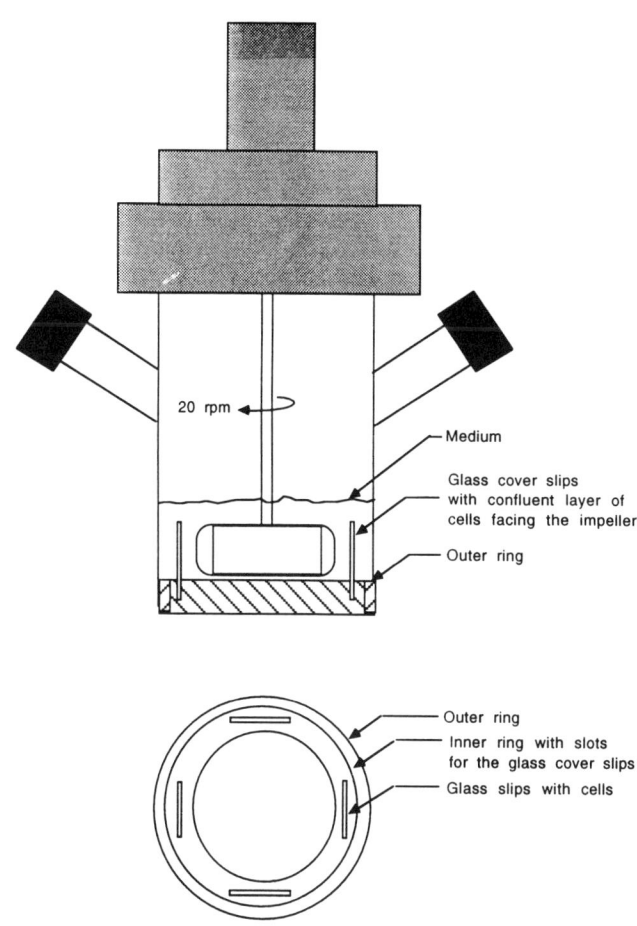

Figure 2. The Spinner flask apparatus

2mM $MgCl_2$, 0.1 % Triton X-100 for 10 min at room temperature (Bacallao and Stelzer [12]). To decrease the autofluorescence of the glutaraldehyde-fixed cells, the cells are then washed three times with 1mg/1ml $NaBH_4$ in phosphate buffered saline with no divalent cations [PBS(-)] for 15 min each at room temperature. This step is repeated for two more times with freshly dissolved $NaBH_4$. The cells are then washed three times for 10 min each with PBS(-). The slides are then washed with PBS(-) containing 0.2% fish skin gelatin (Sigma, FSG) and 0.1% Triton. For the specific detection of F-actin, each monolayer was treated twice with 50 ml rhodamine phalloidin dissolved in PBS(-) containing 0.2% FSG and incubated in a humidified chamber for 45 min. The monolayers were washed twice with PBS(-) plus 0.2% FSG and three times with PBS(-) for 10 min each. The cover slips were placed on the microscope slides with support mounts made out of acrylic nail polish (height < 40 μm). The cover slips are placed on the microscope slide with the cells facing the slide along with a drop of quenching solution (50% glycerol, PBS(-), 1mM NaN_3, 100 mg/ml 1,4 Diazabicyclo(2,2,2)octane).

Microscopy

The slides were observed using confocal microscopy (BioRad MRC-600). A 63X oil immersion objective was used to focus on the basal plane to observe the F-actin stress fibers. Comparative analyses were made on monolayers from the same passage, stained under identical conditions and observed under identical settings to allow a quantitative assessment. Fluorescence intensity of rhodamine-phalloidin-stained F-actin is directly correlated with the F-actin quantity determined by the conventional biochemical method (Southwick et al. [13]). To account for the number of detached cells, the fluorescence values were normalized to the number of cells in the field of measurement. The fluorescence intensity from the shear stressed monolayers were reported as relative fluorescence intensities versus their paired control monolayers.

Shear Stress Measurements

Constant temperature hot-wire anemometry was used to measure shear stresses. A typical sensor (Dantec Measuring Technology) is an electrically heated wire made of tungsten or platinum, about 1mm long and 5 μm in diameter, supported between two needles by arc welding or soldering (Figure 3).

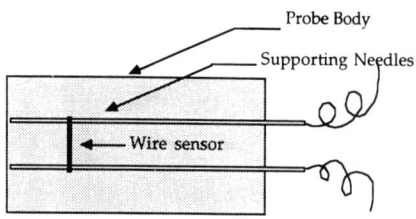

Figure 3: A typical hot wire probe.

The sensor makes one of the arms of the Wheatstone bridge circuit which has two fixed resistors and an adjustable resistor with the differential feedback amplifier sensing the bridge balance and holding the sensor temperature constant. The sensor is placed in the path of the fluid whose shear rate is to be measured. The flowing fluid carries away the heat from the sensor and unbalances the bridge. This causes the feedback amplifier to increase the sensor heating current and to bring the bridge back in balance (Figure 4). Since the feedback amplifier responds rapidly, the sensor temperature remains virtually constant as the velocity changes. The voltage difference across the bridge is proportional to the fluid velocity.

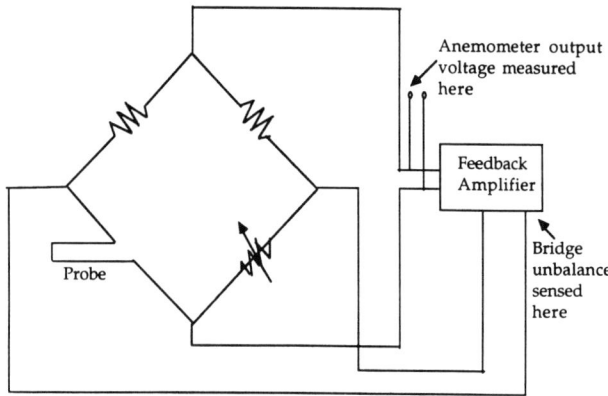

Figure 4: The probe acts as one resistor in the Wheatstone bridge and feedback amplifier adjusts current to maintain the bridge balance.

The probe was initially calibrated using a laminar flow through a 2 foot long, 10 mm square borosilicate glass tubing (Vitro Dynamics Inc.). The calibration curve is obtained by plotting the cube root of the shear stress against the square of the voltage. This relationship is derived by equating heat transfer from the sensor to the fluid during laminar flow and the electrical power input required to maintain the sensor heated resistance (Geremia [14]).

The calibrated probe was then glued on a cover slip and placed in the spinner flask facing the impeller in the same position as the cells. A Macintosh II computer with MacAdios data collection hardware and software was used to measure voltage fluctuations. The small fluctuations were resolved using a high pass R-C circuit removing the mean signal from the input to the MacAdios. An oscilloscope was used to read the mean signal which was then added to the fluctuations before calculating the shear stresses.

ATP Depletion

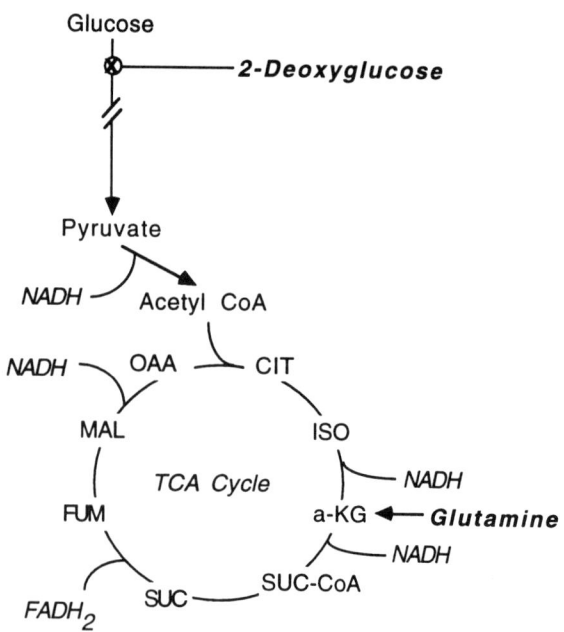

Fig 5: Inhibitors and substrates of glycolysis and the tricarboxylic acid cycle. Glutamine was the only exogenous substrate available during preincubation

and glycolysis was competitively inhibited using 2-deoxyglucose.

MDCK cells produce ATP by both glycolytic and oxidative metabolism. To produce a rapid depletion of ATP, the cells were pretreated in Solution A (Table 1) for 3 hr at 37°C (Mandel et al. [15], Doctor et al. [16]). Solution A, which contains the oxidative substrate glutamine as the sole metabolic substrate, was utilized to deplete the endogenous stores of glycolytic substrates (Figure 5). Following the preincubation period, the monolayers were washed and incubated in Solution B. This solution inhibits glycolytic metabolism by removal of exogenous glucose and the addition of 2-deoxyglucose, a competitive inhibitor of glucose. Oxidative metabolism was inhibited chemically by the addition of 10 μM rotenone, a mitochondrial inhibitor. Monolayers were incubated in solution B for either 0, 5 or 20 min (Figure 6).

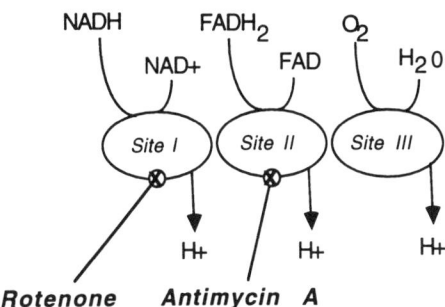

Fig. 6: Inhibition of the mitochondrial electron transport chain by addition of the site I inhibitor rotenone.

ATP and Protein measurements

After the experimental periods, the cells were scraped off the slides into 1.0 ml of ice-cold 6% perchloric acid. The protein was pelleted at 14,000 xg and quantified using a bicinchoninic acid protein assay (Pierce, IL). The supernatant was neutralized with 10% 3.8 M KOH/1.55 M K_2CO_3, the precipitate was pelleted at 14,000 xg and the supernatant was assayed for ATP content by the Luciferin-Liciferase assay. (Stanley and Williams [17]).

TABLE 1. Cell Culture Incubation Solution

	Solution A	Solution B
	PREINCUBATION	ATP DEPLETION
pH	7.4	7.4
	(mM)	(mM)
NaCl	125	125
KH_2PO_4	5	5
$MgSO_4$	2	2
$NaHCO_3$	25	25
$CaCl_2$	1.5	1.5
Glutamine	2	0
Deoxyglucose	0	10
Rotenone	0	10 (μM)
Adenosine	0	1
Allopurinol	0	0.2

Statistics

The multiple samples in the time course studies were subjected to an analysis of variance, with $P<0.05$ used to denote significant difference.

RESULTS AND DISCUSSION

Anemometry

The fluid-imposed hemodynamic forces may be expressed as stress, or force per unit area, and includes both a normal component, pressure and a tangential component, shear stress. Wall shear stress represents the frictional force per unit area and is a function of the local velocity gradient. For a fluid exhibiting Newtonian behavior there is a linear relationship between the shear stress exerted on the wall τ_w and the local velocity gradient at the wall, namely the wall shear rate γ_w, such that

$$\tau_w = \mu (du/dy)_w = \mu \gamma_w$$

where u is the velocity component in the direction parallel to the wall, y is distance in the direction perpendicular to the wall, and μ is the viscosity. The wall shear rate is a direct reflection of the local velocity pattern at the wall.

Hot wire anemometry was used to measure the wall shear rate experienced by the cells inside the spinner flask. The probe was calibrated using laminar flow in a square tube. The shear stress on the wall of the square tube during steady flow was derived from the exact solution for fully-developed laminar flow velocity in a rectangular tube (Lundgren, Sparrow and Starr [18]). A brief calculation follows:
  Length of the side of the square tube = 1.0 cm.
  Radius of the tube = a = side of the square/2
  Viscosity = μ = 0.0075 g/cm s
  Density = r = 1.00 gm/cm^3
  Shear rate = γ (1/s)
      = 1.2 (Flow rate Q)/ (a)^3
  Shear Stress (τ) = (Viscosity) x (Shear rate)
  Velocity V(cm/s) = Flow rate Q/ c. s. area
  Reynolds Number = Re = (a) (V) (ρ) / μ

The flow through a tube is considered steady or laminar if Re < 2100 and turbulent when Re > 4000.

The fluctuations measured from the oscilloscope were added to the mean voltage of 2.26 V when the probe was placed on the glass slide inside the spinner flask and the shear rate was obtained using the calibration curve. The shear stress profile experienced by the cells inside the spinner flask at 20 rpm was plotted over a representative segment of time (Figure 7). 4096 data points were collected over 8 seconds to produce this profile. The shear stress fluctuated from as low as 0.02 to 0.27 dyn/cm^2 and the mean shear stress was seen to be 0.092 dyn/cm^2. The observed peaks occur at every 1.5 s, corresponding to the passage of the impeller tip at 20 rpm.

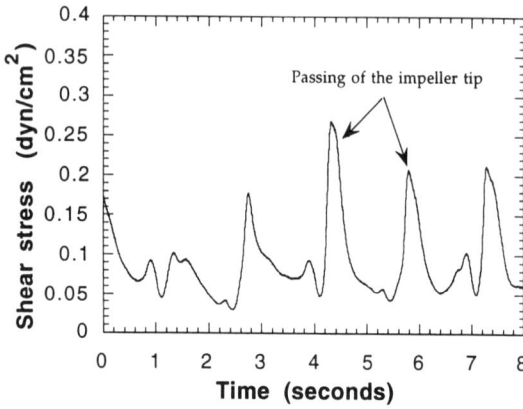

Figure 7: Shear stress profile experienced by the cells for a period of 8 seconds inside the spinner flask at an impeller speed of 20 rpm.

The impeller Reynolds number of 700 showed that the flow in the spinner flask was in the near-turbulent regime. It is suggested by the Kolmogorov theory that the eddies, which are the basic structure of turbulence, of sizes less than the cell diameter cause pressure differences across the cell and deformation of the surface that could cause cell damage or lysis (Cherry and Papoutsakis [7]). The Kolmogorov eddy size in the spinner flask system was calculated to be 250 μm, assuming the maximum energy dissipation to be around the impeller region. Although it is unclear if glass cover slips could breakup these large eddies into smaller ones, comparable to the cell size and capable of dissipating enough energy to cause cell damage, the Kolmogorov eddy size was considerably larger than the cells which are of the size ranging from 10-15 microns in diameter.

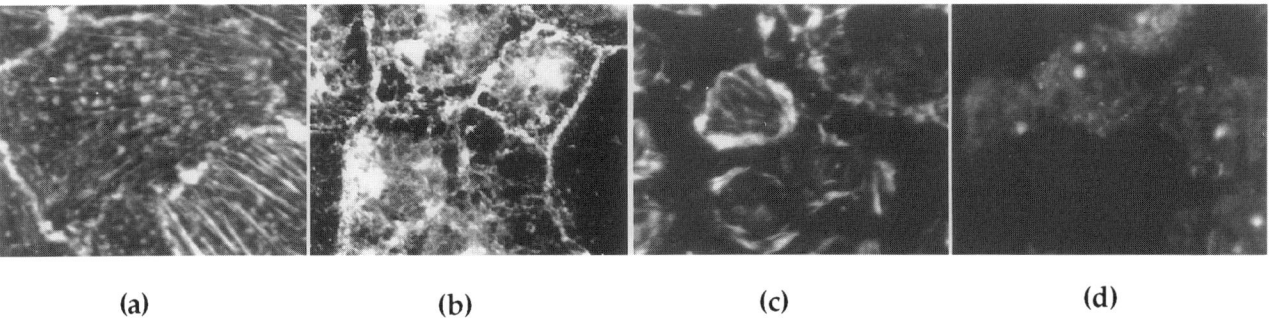

Figure 8 a-d:

(a) The controls (No Stress).
(c) Stressed condition for 6 hrs.

(b) Stressed condition for 3 hrs.
(d) Stressed condition for 12hrs.

F-actin Staining and Quantitation

As previously shown, unstressed MDCK cells displayed brightly stained, well-developed stress fibers in a random orientation along the basal aspect of the cells (Bacallao et al., [10], [11]). The lateral aspects of these cells showed prominent cortical F-actin filaments (Figure 8a). Within 3 hrs of stressed condition, the basal F-actin filament intensity was diminished by 23% and appeared in a more centralized, perinuclear localization (Figures 8b and 9). The cortical F-actin remained largely unchanged. After 6hrs of stress, the cortical F-actin was now diminished, basal F-actin staining intensity was down to 63% of control values and 50% of the cells had detached (Figures 8c and 9). After 12 hr of shear stress, very few cells remained attached to the slide. Among the remaining cells, the F-actin content had diminished to 15% (Figures 8d and 9).

Energy Depletion

Bacallao et al. [11] demonstrated that ATP-depleted MDCK cells underwent a consistent, rapid decrease in stress fiber density and detachment from its substrate. ATP, the metabolic currency within cells, was measured to determine the metabolic status of the cell. Under control corditions, the ATP content was 124 nmoles/mg protein. Inhibition of glycolytic and oxidative metabolism resulted in a 45% decrease in ATP/mg protein. After 20 min of metabolic inhibition, ATP levels declined to 21% of control values (Figure 10). As described previously, these cells responded to the depletion of ATP by depolymerizing the F-actin cytoskeletal network (Bacallao et al. [11]). The initial loss of F-actin occurred in the basal region of the cell followed by losses in the cortical and terminal web regions. Subsequently, after the stress fibers were largely diminished, the amount of cell detachment increased dramatically.

Shear Stress and Cellular Energy Status

The terminal cellular effects of shear stress and ATP depletion, namely F-actin depolymerization and cell detachment, were identical. To determine if the cellular mechanism of shear stress worked through

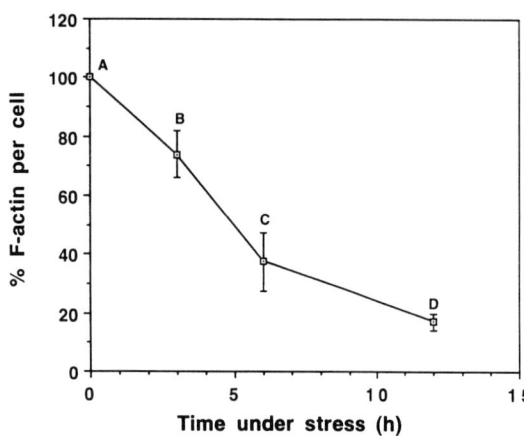

Fig 9: Comparative F-actin content per cell in the basolateral layer. The amount of F-actin decreases with increasing period of time in the stressed condition. Experimental points without a common letter are significantly different from each other (P<0.05).

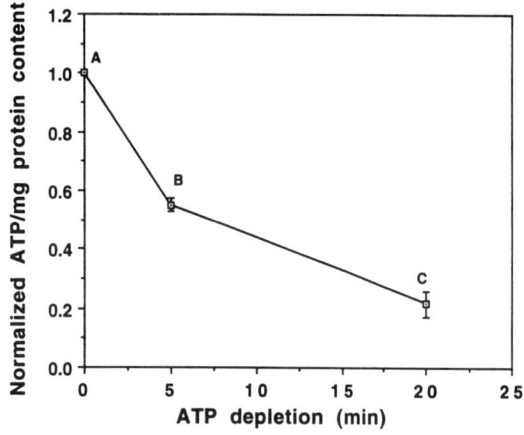

Fig. 10: The ATP/ mg protein decreasing with time, using the inhibition of glycolysis and the mitochondrial electron chain. Experimental points without a common letter are significantly different from each other (P<0.05).

ATP depletion as an intermediary step, cellular ATP was measured during the application of shear stress. Since cell detachment could also occur from direct shear stress, the number of attached cells was assessed and the ATP values were corrected for the concentration of protein (Figure 11). The protein content per cell was stable and independent of the duration of shear stress. There was no statistical difference between the ATP/mg protein for the controls and the stressed condition demonstrating that the ATP content of the cells do not change due to externally applied stresses (Figure 12).

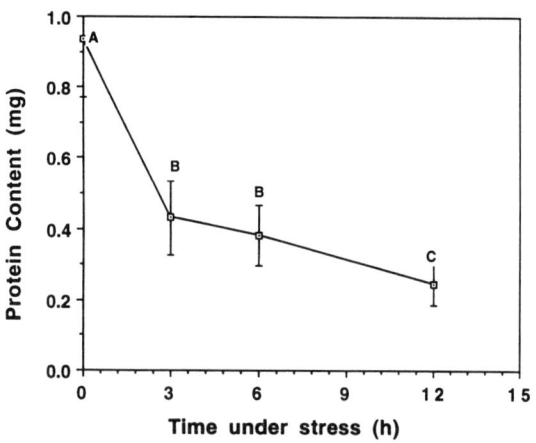

Figure 11: The decrease in the total protein content per slide with increasing exposure to stress. The total protein content decreased as the number of the cells on the slide fell off. Experimental points without a common letter are significantly different from each other (P<0.05).

In summary, MDCK cells were found to be extremely sensitive to shear forces. In less than 12 h, a mean shear stress of 0.09 dynes/cm^2 induced the depolymerization of the actin cytoskeleton and the detachment of cells from the cover slips. While chemical anoxia was confirmed to lead to a similar cytoskeleton breakdown and cellular detachment it was determined that ATP remained unchanged in shear stressed cells. Thus, the cytoskeleton breakdown in shear stressed cells was not due to metabolic deprivation and must occur by a mechanism independent of ATP depletion.

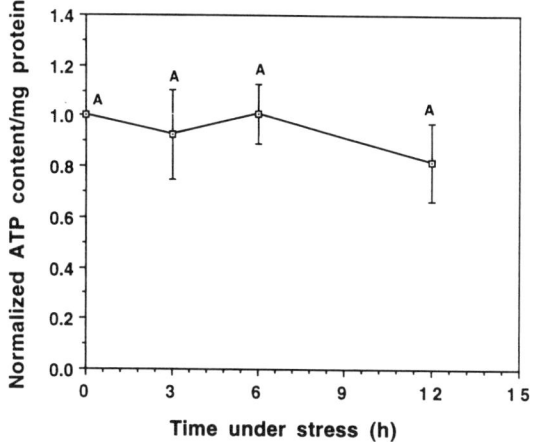

Fig. 12: The normalized ATP/mg of protein values for the control and the stressed conditions. Experimental points without a common letter are significantly different from each other (P<0.05).

Effects of the Cultured Cell's Origin

These results demonstrated that mammalian cell types of different origins have distinct responses to shear stress. Bovine aortic endothelial cells (BAEC's) are seen to be much more shear resistant and can withstand shear stresses upto 5-10 dyn/cm^2 before showing morphological changes (Mo et al. [19]) which is very close to the amount of shear stresses experienced by these cells *in vivo*. In contrast, MDCK renal epithelial cells were very sensitive to shear stress, detaching under only 0.1 dyn/cm^2 of stress. The progenitors of these cells were exposed to the shear stress created by flow of the glomerular filtrate. An approximate calculation for the *in vivo* Reynolds number in the proximal tubules of the kidney is done below:

Total flow of fluid = 125 ml/min
Total number of nephrons = 1.2 million
Diameter of each nephron = D = 25 μm
$= 25 \times 10^{-4}$ cm.

Density of fluid = 1 gm/cm^3.
Viscosity of the fluid = 0.0075 gm/cm.s
Flow rate per nephron = 125/1.2
$= 104$ nl/min $= 1.733$ nl/s
Velocity of the flow = Flow rate/(3.14 x D^2/4)
$$= \frac{1.733 \times 10\exp[-6]}{3.14 \times \frac{(0.0025)^2}{4}}$$
$= 1.109$ cm/s
Reynolds number = Re = Dvρ/μ
$$= \frac{0.0025 \times 1.09 \times 1}{0.0075} = 0.354$$

The proximal tubules experience maximal flow and the above calculations are based on this flow rate. Distal tubules, of which MDCK cells were initally derived, experience a much lower flow rate and extremely low shear stresses. Though a thorough investigation is required, these observations suggest that the original shear stress environment of the cultured cells may dictate their resistance to the effects of shear stress. This may become an important factor when selecting a mammalian cell line for large scale industrial bioreactors.

Cellular Mechanisms of Shear Stress and Cell Detachment

While it is unclear what the cellular mechanism is that leads to the F-actin depolymerization and cell detachment, shear stress can activate ion channels in the plasma membrane (Olesen et al. [20], Lansman et al. [21]). These channel activations could initiate the transduction of shear stress into cellular effects. Ando et al. [22] have shown that flow induced changes in the Ca^{+2} content of vascular endothelial cells. This is of special interest since Ca^{+2} and Ca^{+2}-calmodulin is intricately involved in the regulation of numerous actin binding proteins, bundling proteins, capping proteins and severing proteins. Unraveling the cellular mechanisms responsible for the cellular alterations invoked by shear will remain a key in the future development of efficient and productive mammalian cell bioreactors.

ACKNOWLEDGMENTS

The authors acknowledge the time and patience of Dr. Brian Doctor for his assistance with preparation of the manuscript and many useful suggestions. Partial support for this work was provided by North Carolina Biotechnology Center. Vinayak D. Bhat was supported by Duke University Graduate Fellowship.

LITERATURE CITED

1. Midler, Jr. M. and Finn, R. K.,*Biotechnol. Bioeng.*, **8**, 71-84, (1966).

2. Levesque, M. J. and Nerem, R. M., *J of Biomechanical Engineering*, **107**, 341-347, (1985).

3. Stathopoulos, N. A., and Hellums, J. D., *Biotechnol. Bioeng.*, **27**, 1021-1026, (1985).

4. Ookawa, K., Sato, M., Ohshima, N., *J. Biomechanics*, **25**, No 11, 1321-1328, (1992).

5. Alberts, B., Bray, D., Lewis, J., Raff, M., Roberts, K., Watson, J.; Molecular Biology of The Cell; Third edition, Garland Publishing, Inc., New York & London, (1994).

6. Peterson, J. F., Ph.D. thesis, Rice University, Houston, Texas, U.S.A. (1989).

7. Cherry, R. S. and Papoutsakis, E. T., *Animal Cell. Biotech.*, **4** 71-121, (1990).

8. Dewey, C. F., Bussolari, S. R., Gimbrone, M. A., Davis, P. F., ASME *J. Biomedical Engineering*, **103**, 177-185, (1981).

9. Eskin, S. G., Ives, C. L., McIntire, L. V., Navarro, L. T., *Microvascular Research*, **28**, 87-94, (1984).

10. Bacallao, R., Antony, C., Dotti, C., Karsenti, K., Stelzer, H. K., Simons, K., *J. Cell Biol.*, **109**, 2817-2832, (1989).

11. Bacallao, R.; Garfinkel, A.; Monke, S.; Zampighi, G.; Mandel, L. J., *J. Cell. Sci.*, **107**, 3301-3313, (1994).

12. Bacallao, R. and Stelzer, E. H. K., *Methods in Cell Biol.*, **31**, 437-451, (1989).

13. Southwick, F. S., Dabiri, G. A., Paschetto, M., and Zigmond, S. H.,*J. Cell Biol.*, **109**, 1561-1569, (1989).

14. Geremia, J. O., *DISA Information*, 1972, **13**, 5-10.

15. Mandel, L. J., Bacallao, R., and Zampighi, G., *Nature Lond.*, **361**, 552-555, (1993).

16. Doctor, B. R., Bacallao, R., and Mandel, L. J., *Am. J. Physiol.* 266, Cell Physiol., **35**, (1994).

17. Stanley, P. E. and Williams, S. G., *Anal. Biochem.*, **29**, 381-392, (1969).

18. Lundgren, T. S., Sparrow, E. M., and Starr, J. B., *Trans. ASME: J. Basic Engineering*, 620-626, (1964).

19. Mo, M., Eskin, E. S., Schlling, W. P., *Am. J. Physiol.,* **260**, H1698-H1707, (1991).

20. Olesen, S. P., Clapham, D. E., Davies, P. F., *Nature,* **331**, 168-170, (1988).

21. Lansman, J. B., Hallam, T. J., Rink, T. J.,*Nature,* **325**, 811-813, (1987).

22. Ando, J. T., Komatuda, and Kamiya. A., *In. Vitro Cell. Dev. Biol.,* **24**, 871-877, (1988).

Index

A

Actin cytoskeleton 166
Agitated vessels (mechanically)....131
Analysis
　hydrodynamic 20
　of mixing 1
Average shear stress 146
Axially stirred vessel 102

B

Bubble columns 11

C

Cells, cultured apithelial 166
CFDS-FLOW3D, application of . 150
Chemical reactions 31
Comparison of experimental data.. 39
Conical centrifugal film reactor .. 61
Crystallizers, evaporative 39
Cultured epithelial cells 166

D

Dissipation, power 146

E

Electrostatic spraying of gases . 52
Energy spectrum function 95
Enhanced mass transfer 80
Experimental
　investigation of solids 131
　(LDA) and numerical study ... 102
　verification 11,139
Evaporative crystallizers 39
Experimental data, comparison of...39

F

Film
　flow 70
　laminar rippling 61
　thickness profile 61
　reactor, conical centrifugal 61
Fluids, viscoelastic 115
Flow
　film 70
　semi-direct simulation of 88
　ripple 80
Fluids, viscoelastic 115

Fluidized bed units 45

G

Gases into liquids 52
Gas-liquid
　flow 11
　mixing 161
Grinding in fluidized bed units .. 45

H

High solids loadings 131

J

Jet reactor, tubular 31

L

Laminar rippling film 61
Liquid-solid
　agitation performance 139
　suspension 123
Loadings, high solids 131

M

Mass Transfer
　enhanced, in ripple flow 70
　in a laminar rippling film 61
　in tall tanks 161
　mechanically agitated vessels...131
Micromixing, study of 123
Mixing
　in the Sulzer SMV Mixer 1
　of viscoelastic fluids 115
　simulation of 39
　tank 146
　turbulent 31
　vessels 150

N

Near wall and bottom regions ...102
Numerical
　simulation.............................1,11,39
　study 102

P

Paddle impeller, vessel with 88

PDF modelling 31
Power dissipation 146

R

Reactor
　stirred 123
　tubular jet 31
　two-phase tubular 20
Reynolds number, function of........ 45
Rules from a viewpoint of energy...95

S

Scale-up rule and evaluation ... 95
Semi-direct simulation of flow ... 88
Shear stress
　average 146
　hydrodynamic 166
Simulation of performance 139
Solids suspension 131
Spraying of gases into liquids ... 52
Stirred
　reactor 123
　vessels 102
Sulzer SMV Mixer 1

T

Tall tanks, mass transfer in 161
Thrust force 146
Tubular reactor
　jet 39
　two-phase 20
Turbulent
　flow behavior 102
　mixing 31
　transition in a vessel 88
Two-phase tubular reactor 20

V

Vessel(s)
　mechanically agitated 131
　mixing 150
　with paddle impeller 88
Viscoelastic fluids 115

W

Wall and bottom, near............102,11